STONE ON STONE
The men who built the cathedrals

IMOGEN CORRIGAN

STONE ON STONE
The men who built the cathedrals

ROBERT HALE

First published in 2018 by
Robert Hale, an imprint of
The Crowood Press Ltd,
Ramsbury, Marlborough
Wiltshire SN8 2HR

www.crowood.com

British Library Cataloguing-in-Publication Data
A catalogue record for this book is available from the
British Library.

ISBN 978 0 7198 2798 3

Typeset by Chapter One Book Production, Knebworth

Printed and bound in India by Parksons Graphics

CONTENTS

For Bridget Harrison

According to the grace of God which is given to me, as a wise masterbuilder, I have laid the foundations, and another buildeth thereon.
1 Corinthians 3: 10–15

FIGURES

The following figures are provided within a plate section:

INTRODUCTION

STANDING IN THE nave of Ely cathedral when I was fourteen I became bored with the local guide telling us how long and how high and how many tiles, and I started wondering how on earth ordinary human beings could have created such sky-scraping, dizzyingly high buildings on which even the uppermost parts were delicately decorated. 'How?', 'why?' and 'for whom?' were my unanswered questions. For God, for personal redemption and to make a living are the answers to the last two questions, but the 'how' was harder to fathom. I set out to find the Master Masons – the men who both designed the buildings and ran the construction site. They commanded everything, whether it was sourcing vast quantities of wood and stone, recruiting the workforce, working out the budget, or having enough knowledge about the numerous trades on site to be able to create Heaven on earth out of the cacophony of thousands of chisels and hammers.

Their aim was to do exactly that: the churches were not just for the glory of God, but so that the medieval man and woman would have a better understanding of Paradise through being able to see it on earth. The Master Masons saw themselves as building the City of Heaven or as reconstructing Solomon's Temple, which was a privilege, but also a responsibility. As time went on, men who could create such structures took on a near superstar status. Our ancestors have been criticized (and were at the time, too) for spending huge sums when there was so much need elsewhere, but to take that line is to ignore the fact that they were also fulfilling a genuine need to get closer to God and to the saints. It was the saints and the Church in general who got people through what was often a daily grind. They offered hope and protection and, if you were really lucky, a miraculous cure from illness. The churches and cathedrals offered physical shelter, mystique and beauty; another way to take your mind off what could be the grim reality of daily life. As you read this book, always keep in your mind that the population believed in God and the next life, and that their daily lives were shaped by the Church calendar. Sadly, this did not stop them from misbehaving – even Master Masons fell from grace occasionally, as we shall see. Also, bear in mind that people could repent either in the sense of Confession or by donations, and some gave significant amounts to church building as a way of salving their consciences. Happily, most led God-fearing lives, but I always point out that our ancestors are just like us: they were intelligent, intellectual, optimistic, miserable, sympathetic and jealous, generous and greedy. They shared our hopes and fears for their children and their own future, they had a sharp sense of humour and they wanted to live in comfort. They simply danced to a different beat and that was the beat of Christianity, which both awed and sustained them.

It is an astonishing fact that many of our greatest cathedrals were built against a backdrop of plague, other persistent diseases and warfare. It also surprises people today that Master Masons were willing to embark on projects that they knew could not possibly be finished in their own lifetime. On the other hand, the period we call the Middle Ages was a time of innovation and experimentation with new artistic and architectural ideas, all of which Master Masons seem to have embraced. We can see the results, but not always the making: remarkably few manuals have survived, if they were ever written. Apprentice training was rigorous and no doubt repetitive. It became highly regulated so that only the most competent made it through to their basic trade, and most would have been content to stay at that level. No one began as a Master Mason. They all started their professional life on the lowest rung of the ladder, so to speak, training and working as stone-cutters, masons or carvers. Only a particular few made it to the status of Master Mason, and they were the ones who were not only especially talented craftsmen, but who also proved to be charismatic leaders.

That said, they were continually checked by their fellow Master Masons in the interests of making a building as strong as possible: we only see their successes, after all. They were real people who got into trouble with the law, who occasionally cheated on contracts, who liked to start a job but not to finish it. Some of them were highly litigious (always a blessing to future historians) and some took their place amongst the great of the land. It was unusual for them not to become wealthy and many were famous in their own lifetime even, occasionally, having the privilege of being buried in a prestigious place within the cathedral they were working on. We know the names of hundreds of Master Masons (and sometimes of their workforce, too), but there must have been many more who have disappeared into the blur of history.

The question of how they did it cannot be answered in full, but I hope that this book goes some way towards realizing the hopes and hurdles they had to cross, the kind of issues that they had to take into account, and how they overcame problems as they built. Their creations remain to this day; some breathtakingly beautiful in their exquisite detail, causing us, centuries later, to stand and wonder.

I would like to thank all those who have helped along the way and are too many to list by name, whether they have pointed me in the right direction or allowed me to run free in their church. I visited most English cathedrals and many European ones in the preparation of this book. I was impressed again and again by the welcome received, from Durham to Canterbury and everywhere in between, so my heartfelt thanks to those who stopped to answer foolish questions from yet another visitor. I am fortunate to be a lecturer on Anglo-Saxon and medieval history and art, and am keenly aware of the support I have had from individuals in audiences whose sharp questions have sometimes given me a new line of enquiry to follow. I am sorry that I do not know your names. Likewise, there has been massive anonymous support in the form of hundreds of churches left open for passers-by to visit; this is worth more than rubies to struggling researchers, as are the legions of volunteers working in great cathedrals and modest parishes, giving out leaflets and welcome in equal measure.

I am grateful to all those who have laboured to translate documents to enable me to read these so easily, and I am most particularly indebted to the late Dr John Harvey whose biographical dictionary of Master Masons has been the single most valuable source, often pointing me in the direction of more detailed information. I am older and wiser at the end of the project, but delighted to have been able to do it. Nowadays, it is common to describe any undertaking as a real journey, but this really has been one such: I have travelled miles and only wish that my understanding had grown in proportion. Mine eyes have truly seen some glories. Lastly, but most importantly, I have been especially lucky in the continual encouragement, friendship and positive suggestions from Winston B. Brown, Jackie Cooke, Bridget Harrison, Dr Luella Hibous, Pat May, Caroline Stapleton, Dr Sheila Sweetinburgh, and most particularly from David Spenceley, not to mention my long-suffering and faintly gob-smacked heroic husband, Gordon, who might not mind if he never hears about a cathedral ever again.

CHAPTER ONE

NEW TECHNOLOGY OR NEW SPIRITUALITY?

Establishing the basic shape of a cathedral

BEFORE WE MEET the Master Masons themselves, we need to think about what was at the centre of their being: the cathedral. More especially, we need to consider how the shape of the building developed, which was, after all, critical to the overall plan. In cathedrals and churches, the shape is more important than it might first seem because this affected the spread of the religion. While it is obvious that there cannot be too many variations in the shape of any structure required for public gatherings, the Roman basilica's internal floor plan was suited to Christian meetings because it was essentially an oblong hall with a rounded apse at the most significant part, for it was within the apse that an altar could be placed.

Figures 1 and 2 show basic layouts for both a basilica and a later Gothic cathedral, and it is plain to see how complex the central design became. Early missionaries as far back as the fourth and fifth centuries AD discovered that a large narthex, or porch attached to the basilica, was an important factor and a useful aid to recruiting: anyone could go inside to shelter from the elements, conduct business or simply meet friends. While they were there, they would be able to hear the strangely soothing and seductive mantras of the liturgy being carried out and, no doubt, smell the incense used more liberally then than now. There would normally be three doorways from the porch into the church through which the curious could stare, although the uninitiated would not be permitted to go into the main part of the building, which was reserved for those fortunate enough to have been saved spiritually. One might imagine the craning of necks and whispering in the outer porch; Christianity was new to many and therefore either exciting or perhaps horrifying. It would be natural for many to feel extremely uneasy about this new religion, as anyone would if asked to discard whatever 13

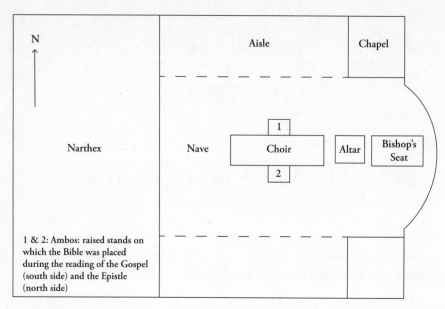

1 Ground plan for Roman basilica

2 Ground plan for cathedral

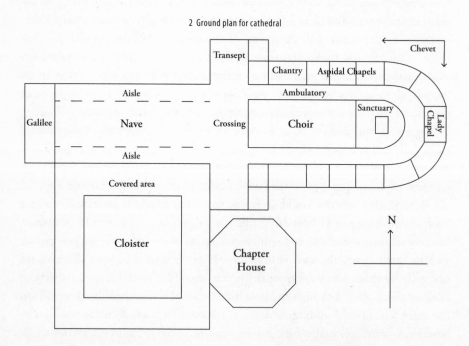

spiritual practice and belief had been ingrained from childhood. By the fourth century AD, Christianity was now seen as a definite religion as opposed to a group of people following the teachings of a charismatic speaker. Enough people had died for it to make it both interesting and credible. The Emperor Constantine's embracing it gave it authority and status, and promises of eternal life and/or relief from physical or mental pain had to make it worth a second look.

Rumours will have circulated in the porch about miracles happening in the name of Jesus Christ and the saints. Local Christian teachers would have sat there, talking to passers-by and the interested and thus this square, almost empty space became a valuable part of the conversion process. In parish churches later, the porch would become the place where civil business was conducted as well as marriage services. It became a space for business for the local community, which is why some later porches are very large, having benches and often niches for statues and holy water to be used in the swearing of oaths. Some porches still have an upper room, which was used for parish meetings and schooling. Given that illustrations have always been an important part of missionary work, stories from the life of Christ or key elements of Christianity would have been painted on the walls or carved over the doorways.

Baptism was the next step. In the ruins of the early Christian basilica at Soli in North Cyprus, which was built in the second half of the fourth century AD, one can see the remains of the baptizing pool just inside the church, immediately beyond the door on the southern side of the porch. In effect, no one could get past without having been admitted to Christianity. To this day, the font is often still placed so close to the main entrance of the church that it is almost an obstacle, a constant reminder of the beginning of the Christian journey, although many congregations have now moved the font to make the baptizing area more central. Once inside the inner building, the newly converted could stand with the others in the areas we know as side aisles. In missionary churches of this style, the aisles would not be open plan and marked with pillars, as they are now, but physically separated by a low wall or fence so that ordinary Christians could see and join in, but not enter. It seems that the congregation could approach as far as the choir area to watch, making the place similar to a theatre with a protruding stage on which the priests would perform. This, again, was an important tool for conversion. There was little enough entertainment for the majority anyway, so the ritual carried out against a backdrop of candlelight, with precious vessels and vestments glinting gold and gorgeous manuscripts glimpsed through a haze of incense, would have been extremely impressive.

In the Roman pagan administrative hall (the basilica), there was often a small room known as a *porticus*, which was accessed from the inside. Sometimes there were two projecting from each side of the building about two-thirds of the way along. When built for Christian purposes, these became small chapels, or even offices, as can be seen marked out on the ground, for example, beside the remains of the seventh-century church at Bradwell-juxta-Mare on the Essex coast. Much later on, these would be extended into the arms known as transepts, which transform the ground plan of the building into the shape of the cross. Gradually, usage and changes in architectural styles would alter the basic floor plan, but the basilica shape appears to have been an effective starting point. St Augustine of Canterbury, travelling from Rome at the end of the sixth century AD, would have been familiar with the basilica-style layout, although we do not know if he intended to impose it on England. Early Anglo-Saxon churches, especially those in the north of England such as Escomb and the older parts of Monkwearmouth and Jarrow reveal a preference for a narrow, single-celled building. These northern churches are a useful indication as to how things might have been because of the influence of Benedict Biscop, who accompanied Theodore, the incoming Archbishop of Canterbury, from Rome in 668–9, worked with him for a couple of years, and then returned to the north-east where he founded the monasteries of Monkwearmouth and Jarrow. The designs are often disproportionately tall, with steeply pointed roofs almost as though they are an arrow pointing upwards, which might have been part of the plan.

Winning hearts, minds and souls

There is, therefore, no doubt that buildings were used as tools for conversion. They had to be impressive to send out the straightforward message that the Christian God was greater than any other gods. We should remember that what look to us to be relatively small-scale stone buildings would have been more striking at a time when most buildings were constructed out of wood or other organic materials. The great builders in stone – the Romans – had left Britain at the beginning of the fifth century AD so their buildings had by now either fallen into decay or been recycled into town walls. The Anglo-Saxon inclination was to build in wood so, while we look at the tenth-century stone church of St Lawrence at Bradford-upon-Avon in Wiltshire, and marvel at its survival, medieval people would have just looked and marvelled. These comparatively small stone churches would also have towered above the wooden, thatched dwellings of the Anglo-Saxons and most would have been visible from some way off. They would have been something so utterly different on the landscape that the simple fact of their presence would have been remarkable to the passer-by. In modern parlance, the 'wow' factor presented by Anglo-Saxon churches is something difficult to imagine in today's steel-and-concrete built environment. The desire of early Anglo-Saxon

Master Masons (and records of a few have come down to us) was not simply to build to the glory of God, but also to provide a roof below which conversion and Christianity could take place – but first the missionary priests had to encourage people to gather below that roof.

It seems likely that the earliest church buildings would have been wooden lean-to arrangements, probably constructed by the priests themselves with local help. But, as Christianity took hold across Britain and Europe, church building progressed from a form of frontier outposts to more impressive monuments to God. The larger ones could attempt to instruct the masses and encourage the priests in a more distinctive way: they could try to recreate Heaven on earth, and this is important. Not only is the Bible peppered with building metaphors, but also with numerous allusions to the Heavenly City. In addition, the Old Testament offers specific details about temple building.[1] The Master Masons did not use these references as any form of template – they were too vague – but they will have noted that there are allusions to structured, planned places in the after-life. This was most notable in the New Testament book of Revelations 3 and, especially, Revelations 21, which was often taken as the authority to lavish fortunes on the decoration of cathedrals and churches. Again, a cathedral was not necessarily built to the specifications laid down, but by the time of the eleventh century there was a great desire to get physically closer to God and to try to understand some of the immense mystery surrounding Him. God was seen as being all-powerful, yet also extremely personal: all sins were known and noted. The risk of damnation was great, but the chance of salvation was also high if one took the right steps. There was much to play for.

The early missionary bishops did not need a huge building so much as an impressive one because such a building was visual propaganda. In England, we tend to think that a cathedral should be vast in size, but this has not always been the case. The building that holds the cathedra (the bishop's seat) is the cathedral, but the building does not have to be much bigger than a large parish church; size is not always important. In the south of France, for example, there are several delightful cathedrals, such as the one at Lescar, which are not cathedral-sized according to usual expectation. One is not overwhelmed and awed by the sight of Lescar's cathedral, either inside or out, but one is conscious of being in a church of status and one is delighted by the capital carvings and the unusual mosaics.

Why did building styles evolve?

The basic shape for the Christian building was now established, but there were regional variations – indeed, there still are. How did the building of a cathedral evolve to take the form we associate with the Middle Ages? How and why did it change from solid Romanesque to soaring Gothic? As can be seen at Caen, in Normandy, and at Salisbury, in Wiltshire, there is a marked difference between the plainer-seeming, strong-looking Romanesque of the Abbaye aux Dames in Caen (*see* plate 1) and the more ethereal, highly decorated Gothic of Salisbury cathedral (*see* plate 2). The round-arched, strongly built church form had endured as a pattern for centuries and was to continue to do so in southern Europe and Byzantium. Why then was there such a marked changed in style in western and northern Europe, and how did it become so popular so quickly? Was it the result of an advance in technology, a change in artistic taste or some difference in the expression of spirituality? All three is the most likely answer, although the last two factors are the hardest to quantify and were probably linked, since the form of the buildings expressed not just the desire of the heart but were a critical part of the never-ending quest of the soul. As to why the new style was so attractive to the north and west of Europe and less so to the south, there is no obvious answer other than that different styles appeal to different people. The thicker walls and smaller windows of Romanesque architecture keep the interior cooler than does Gothic.

Middle Eastern influences

To consider the technical side of cathedral building first, we know that the Middle Ages saw just as much innovation as any other period in history, but that at this time great advances were being made in the building trades. This was partly because there was so much building work taking place. Consequently, Master Masons had more opportunities to experiment with different decorative ideas and local materials, and to compare work happening on one site with work on others, as well as being able to exchange ideas with journeymen from all over Europe. Expertises (of which more in Chapter 5) and other regulatory bodies, such as guilds, certainly helped to spread ideas and improve working practices throughout the industry. Some of the new machinery and ideas appear to have come from the Middle East and are presumed to have arrived in Britain with returning pilgrims and Crusaders.

This notion that Crusaders came back with a new approach to building is reinforced by some of the distinctly Arabic shapes that became so much a feature of Gothic architectural decoration: octagons, hexagons, quatrefoils, ogee domes and geometric patterns are all reminiscent of building in Arabia or Moorish Spain. Also in favour of the Crusaders as a conduit for technology is the timing of the change from Romanesque to Gothic. The Moors, or Berber tribes, had been in Spain since AD 711, so one might have expected to see their artistic influence earlier than the twelfth century if the Gothic style had been inspired by them. The First Crusade had been proclaimed in November 1095 by Pope Urban II to protect pilgrimage routes to the Holy Land and Christian holy places, then under threat from Seljuk conquests, and the campaign culminated in the Fall of Jerusalem in July 1099. At the very beginning of the twelfth century, the first vaulted roof seen in Western Europe was constructed at Durham cathedral, which has some of the finest Romanesque architecture anywhere. The roof could have been due to the influence of newly gained Middle Eastern knowledge, or even the arrival locally of a Saracen mason taken as prisoner. We know that this was the fate of at least one man who came to be known as 'Lalys'. He was captured and taken to South Wales where he made his name as a Master Mason and was said eventually to have been employed by Henry I.[2]

All that being so, one problem with aligning artistic changes with the Crusades is that then one might expect more copies of the ultimate Christian site – the Church of the Holy Sepulchre in Jerusalem. That church's rotunda must have made a strong impression and, indeed, most round churches are associated with the Knights Templar. Some, though, are most definitely not: for example, the tiny round chapel at Lanleff in Brittany, where a notice states emphatically that no man from that village ever went on crusade (no reason is given why). The counter-argument is that the round choir is, indeed, seen in the Gothic cathedral, but taking the form of a rounded end that grows from the body of the church. However, this does not stand up to scrutiny because, as mentioned, churches have long had a rounded apse as a hangover from the shape of a Roman basilica. The rotunda itself is seen in England in the ruins of St Augustine's Abbey at Canterbury. This was begun by Abbot Wulfric II in 1050 (though never completed) – a good half-century before any crusade, although people from Britain had travelled to the Holy Land before that (the earliest recorded English pilgrim to the Holy Land was St Willibald in AD 722). The rebuilding of the Church of the Holy Sepulchre also dates to the middle of the eleventh century.

But the evidence of Middle Eastern influences cannot be denied. Today, were we to design a building, we would expect to draw it out using measurements in number format. This would be so elementary that I doubt many of us would think twice about it, but such thinking was less obvious – indeed, not obvious at all – to the early medieval Master Masons. They used Roman numerals and went on doing so until surprisingly late in history. The Roman system, which substitutes letters for numbers, does work and is still used to this day (not least, to distinguish between kings and other rulers). We would not, however, dream of trying to use the Roman system to add or subtract, and with Roman numerals multiplication and division seem difficult beyond belief. To work out problems such as how long it was between the Battles of Hastings and Bosworth, one would have to subtract MLXVI from MCDLXXXV. The answer, of course, is CDXIX. It is all the more remarkable that the Romans were such great builders themselves, proving that such a system of numbering, though unwieldy, was not impossible to use.

A number sequence using Arabic numerals appears to have been started in India and been used in the Middle East as early as the sixth century. Given that trade routes between Europe, India and Africa had been established by the Romans and given that they were not afraid to embrace new technology, it is odd that the Romans themselves did not adopt the numerals they found there (not least because they did use an abacus for quick reckoning). The first surviving record of Arabic numerals in the West (but only the numbers one to nine, not zero) has been found in the *Codex Vigilanus*, which was compiled in AD 976, suggesting that the Indian-Arabic method was introduced into Spain *c.*900 but, again, does not seem to have spread widely beyond this. These numerals cannot have been extensively used throughout the Arab lands because when a Persian engineer and mathematician of the early eleventh century, al-Karkhi, wrote several treatises on calculation he frequently wrote out the numbers as words, presumably to make it clear to his audience.

One of the many changes that occurred in the twelfth century was that serious medical knowledge arrived in Europe from the Middle East. Much of that knowledge had, in turn, originated from ancient Greece. Although medical usage of herbs and so on was highly developed in Europe, attempts to gain more technical knowledge had been regarded with suspicion. This was a time when Christianity taught that the soul was more important than the body and anyone going to help someone who was sick or injured would be advised to send for the priest and to ensure that he had priority in the sick-room. Indeed, that was the ruling laid down by the fourth Lateran Council in 1215. If someone really needed medical help, they were advised to choose the appropriate saint and ask him or her for relief, or even for a miracle; it was part of how both the Church and society operated. Arabic and Jewish peoples thought differently, with the result that their scientific expertise was streets ahead of that in Western Europe. When it came to mathematics, science and medicine, the thinkers of the Middle East were by far and away the leaders in what was then the known world; indeed, words such as algebra, alchemy, alkaline and alcohol have all come to us from Arabic. So, it is not hard to understand that, in the more constructive climate of learning of the Middle East, thinking about the technical aspects of everything also flourished, and this included building and all types of machinery.

Some examples of building machinery influenced by the Middle East

It is hard to quantify what this machinery might precisely have been, although for building it seems likely that it included more sophisticated devices for lifting, measuring and cutting. Sadly, the chroniclers of the time failed to mention the minutiae of a building site. Inventories and other such useful documents are few and far between. The windlass was known in Europe around the year AD 600 since it was described in the last book of Isidore of Seville's *Etymologiae* (also referred to as the encyclopaedia of all things known). The windlass was a useful, albeit unwieldy, apparatus, but probably not as miserable to operate as the human-driven treadmills that go back to Roman times, if not before. There is an image of a manned treadmill on a first-century tomb of the Haterii, now in the Vatican. They were effective because it has been calculated that one man working inside an 8ft-diameter wheel could lift a weight of about twelve hundred pounds, but such wheels frequently caused injury to the human 'hamsters' who operated them.[3] Nonetheless, they appear in numerous manuscript illustrations well into the sixteenth century, so they were evidently the lifting machine of choice; some illustrations show numerous tread-wheels of various sizes at all levels of the building under construction. This type of machinery was still in use in the nineteenth century when John Stuart Mill described them as 'unequalled in the modern annals of legalised torture'. It was often easier to leave them in situ than to dismantle them and, anyway, lifting machines might later be needed for repairs, so occasionally treadmill-cranes can be seen today, for example, in Salisbury and Canterbury cathedrals.

In about 1475 William Orchard paid the abbot of Rewley 10s for a 'great instrument called a crane to lift stones and mortar high up over the wall', and in 1482 William Prentice was paid for repairing a crane at Cambridge.[4] These are late-medieval references and only available to us through surviving financial accounts, but it does not mean that cranes were not in use in Britain earlier than that. A system of counter-balancing weights and pulleys would have been a significant advance on the building site, just to get materials moved about quickly, and this may be the type of technology that returning Crusaders brought home with them.

Styles of vaulting

Rib vaulting is one of the distinctive features of Gothic architecture that is so much a part of our cathedrals in Britain. Quadripartite vaulting (ceilings divided into four sections) is the most usual sort, and frequently found in parish churches as well. When six-part or sexpartite, vaulting is used as, for example, at Laon in northern France, it can produce a wonderful decorative effect because the ribs can be extended down into alternating columns that might be of different sizes, which adds to the overall rhythm of the building. As confidence in vaulting grew, smaller and purely decorative ribs were added, known as tiercerons and liernes. These could be used to make increasingly intricate patterns and, interspersed with the roof bosses especially beloved of British cathedral builders, could make the ceiling a sea of dazzling art in itself. There were some things that the British Master Masons made their very own and were you to wake up and find yourself in a strange cathedral, one way of establishing that you were on home ground, at least, would be to look at the roof. In Britain, roof bosses very often form a row of studs right down the centre of the nave and perhaps into the choir as well – rather like a zip fastener. This has been done with marvellous results at Exeter, Hereford, Lincoln and Norwich cathedrals, to mention but a few. The overall effect helps the eye understand the length of the building, but each boss has its own design. These are sometimes just leafy designs, but they often tell biblical stories or show iconographic symbols. They are an art form in themselves, many cathedrals providing a large, mobile mirror so that they can be examined. The British Master Masons then went a step further and produced fan vaulting, one of the best examples being in the cloisters of Gloucester cathedral. Again, this is rarely seen on the Continent, but is also found in some of Britain's more prosperous parish churches, such as in Ottery St Mary in East Devon or in the Wilcote Chapel of St Mary's, North Leigh, in Oxfordshire.

The reason this could be done so effectively was because rib vaulting distributes the weight differently from a standard flat or barrel roof. Just as a rounded arch is a stronger structure than a flat roof resting on two or more uprights, so a pointed arch can dispense weight downwards. If you add buttressing to stop the weight from pushing the walls outwards, you can make it so that not all of the roof is required to be a solid mass, but can be merely decorative. There is nearly always a gap between the ceiling and the outer roof and, therefore, one could argue that it is the construction of the roof that is the critical element in the shaping of the new-look cathedrals, but it might be helpful to consider just what the specific differences are between Romanesque and Gothic architecture.

The features of Romanesque and Gothic architecture

In England, the Anglo-Saxon church designers had gone for the tall, narrow, single-cell buildings with the rounded apse already mentioned, while a sturdier form was adopted on the other side of the English Channel. After the Conquest, that Continental style was introduced in Britain, where it is known as 'Norman'. More universally, it is known as 'Romanesque', but this tag was not allocated until the 1830s when critics intended it as an insult to those designers who copied the Roman style of building. In effect, it *was* like Roman, but was not as good, being a caricature of their great work. This was then gradually applied to any building – not just nineteenth-century ones – that were in the Roman style, but not actually built by the Romans. This may sound surprising to us, living as we do in an age where anything new in art terms is greeted with applause, even if it is hard to see where the skill lies. In nineteenth- and early twentieth-century France and England, the art critic reigned supreme, making it all but impossible for someone with new ideas to get their work displayed, as the Pre-Raphaelites, and later the Impressionists, were to discover. In a similar way, the term 'Gothic' was intended as an insult, although one that dates back to the 1640s. The Goths were synonymous with barbarians, the Visigoths and Ostrogoths being aggressive peoples not famous for working in harmony with their neighbours. A Visigothic army under Alaric I eventually moved into Italy and sacked Rome in AD 410. It seems a little excessive now to associate the violence of those times with some of our greatest architecture.

Romanesque has rounded arches between immense columns, and small windows set into very thick walls. Gothic has pointed arches, visible buttresses and expanses of glass set in thinner walls. The art of sculptures and manuscripts associated with Romanesque is stylized and often angular with etiolated figures, but also very often made with a light touch, in contrast to the buildings. Gothic sculptural style is softer, much more natural to look at; its statues have far more realistic facial expressions. Whereas Romanesque/Norman architecture had tended to have smaller stones, roughly cut, and roofs supported by immense columns, Gothic had larger stones of smoother dressed ashlar.[5] This again reinforces the idea that some form of lifting device had been discovered or developed, since blocks had been cut smaller before (presumably to make them easier to move around the building site) and early manuscript illustrations show stones being carried by one man in a hod, or two men on a stretcher. The north transept of Winchester cathedral offers us useful evidence here because it has two rounded Romanesque arches, side by side, but one dates to about 1080 and the other to 1108, following the need to make repairs. There is a discernible difference, not just in the size of the stones, but in the quality of their cut. New techniques meant that more accurately cut and smoother-finished stones could be laid in regular courses, giving an even face and finer joints. This gradually did away with the need for bulky columns, which were often hollow in the middle and filled with rubble cores, and allowed walls and shafts to become thin and slender. Unfortunately, over the centuries, the rubble filling in the columns has often packed down, leaving the support for the rest of the building considerably less substantial than the Master Mason had originally intended; as a result, these have become a grave source of concern and fund-raising today.

The obvious stylistic change that Gothic brought was the vanquishing of these mammoth columns, some so stout that their diameter is equal to their height. They were often deeply incised with chevron and other patterns, and are works of art in themselves still visible in the cathedrals of Durham, Norwich, Bayeux, and in Waltham Abbey. These robust-looking columns, which held up lower roofs, were replaced by slender-seeming pillars, sometimes grouped together in bundles soaring to roofs of a greater height. Sometimes this is a trick of the eye, of course; they are rarely individual slender pillars, but do, in fact, comprise one single mighty column cut to create that effect. Very often a Romanesque nave has had a Gothic layer added to make up its height, so both sorts of architecture can be seen at once. Buttressing has been mentioned. This characteristically Gothic feature is essential to act as a brace, holding the structure of the building together, and it arouses mixed feelings: some say that it makes the cathedral look as though it is suspended between spokes and adds to its air of fragility, while others may think that it looks like a gigantic spider crossed with a Meccano set. Either way, when we look at the buttressing we are seeing the workings of the building; there is nothing discreet about this feature, but they become part of the art form at Paris, Rheims, Le Mans, Amiens and numerous others. The spokes of the buttresses often have fancy shapes or tracery cut into them, partly no doubt to help them stand up to strong winds. For example, at Amiens, the buttress tracery is appropriately shaped like small windows (*see* plate 3). They might also have carvings added, ferocious beasts perhaps, to guard the holy end of the church. The progression of this type of Gothic art can be tracked from the plain yet functional buttressing that makes up the chevet (rounded end) of Vézelay's choir, built *c*.1165, to the riot of tiny turrets and *fleurs-de-lys* that adorn Rheims (1211–1241) and Cologne (1248–1322).

The new-style roofs and buttressing and improved lifting devices combined so that, in general, Gothic cathedrals are significantly taller than Romanesque, not counting any spires that may have been added. In France, in particular, they built higher and higher with mixed results, as will be discussed in Chapter 5, the internal height of Beauvais cathedral being over 48m (157ft). The staggering height of these buildings (even more staggering to our medieval ancestors than to us) was very much part of the spirituality of the time. As we have already seen, there was a desire to create Heaven on earth; at the same time, why not try to reach physically up to it as well? Everything, from the sharpened arches to the pinnacle decorations and spires pointing upwards, was there to remind our ancestors of the way their miserable earthly efforts should be directed. Perhaps the English were more prosaic? In England, at least, cathedrals are noted for their great length more than their height.

The outstanding feature of the new Gothic style was glass. Because there was now so much less weight on the walls, there was also far more scope for the walls to be decorated in a way that had never been seen before. Coloured window glass had been used for centuries and, indeed, had been introduced to Britain c.680 when Benedict Biscop installed it at Jarrow, but it had never been possible to have acres of it. In Romanesque churches, windows had been smaller, so the buildings were naturally much darker inside. Interior decoration depended on murals and tapestries, many of which probably could not be seen properly in the gloom. Large numbers of candles were deployed which, when combined with the tapestries and wooden roofs, spelt the death-knell of many Romanesque buildings. The effect of expanses of glass on the inside of the building was dramatic, and many felt that it was the most significant advance not just in architecture, but in gaining an insight into Heaven.

Abbot Suger and his architectural experiment

The man who is generally held to have been responsible for it all was Abbot Suger, who lived from about 1081 to 1151 and who was in charge of the abbey of St-Denis, then a separate entity from Paris, but now part of its northern suburbs. He was a most extraordinary man and so were his works. He was not the oldest son of his family and so had been given to the abbey as a child-oblate; he identified with that place so strongly that he felt it to be part of his very existence. At the time, the buildings had become very run-down and the abbey was seen as a sink of iniquity. It had once been renowned for its wonderful artistic works, but all that had fallen into decay and no one seemed interested in reviving its once great standard of education. This must have seemed more than strange to the young Suger because the abbey had also been a place of coronation and burial of kings for some time, and Suger grew up in the presence of their tombs; men such as Dagobert I, who had died as far back as AD 638. Indeed, the great emperor Charlemagne's grandfather, Charles Martel, had such a devotion to St-Denis that he had been buried in a previous church on or near the site in AD 741. Although it had fallen into a period of decline in terms of fabric, affection and spiritual fashion, it had a sound enough pedigree to be brought back into the centre of royal and noble patronage, and Suger was just the man to do it.

He was a monk, of course, but he also had one of the finest legal minds of the time. He mastered fluent spoken Latin at a young age, which made him extremely useful, Latin being the *lingua franca* of Europe and beyond, and, as he had a real gift for negotiation, he was often employed by the king on foreign affairs. He was physically courageous in warfare (although you might ask what business a monk had on the battlefield), and we know that Suger played an active part in the siege of Le Puiset in 1111. His involvement and blood-lust weighed on his conscience towards the end of his life when he must have wondered how he would pass them off to the Almighty on the Day of Judgement. He, who had been so courageous in active life, wept with fear on his deathbed.

Abbot Suger's best-known legacy was the launching of the Gothic movement, and yet his most high-profile work in his lifetime was as co-Regent of France for two years with Ralph of Vermandois and Archbishop Samson of Rheims, when Louis VII decided to go on Crusade in 1147. The appointment was a remarkable accolade for a man who, while not of common stock, was far from being a noble-man. Being an abbot entailed much responsibility and power, so Suger was not thrust into the limelight from total obscurity; the abbey of St-Denis may have been dilapidated, but it still had high status. Suger was well known at court, regarded as a friend by both Louis VI and Louis VII while having his own set of enemies and detractors, too. He was known to be a strict disciplinarian, ruling the abbey with a firm hand, but he was genuinely concerned about the welfare of his monks and, although he shamelessly favoured his relatives whenever he had the power to do so, he does not appear to have milked the system for himself. He was described as seeming to be impervious either to great sadness or great happiness, and so we might see him as something of a cold fish until we also hear that he loved Christmas and liked telling stories late into the night to the enjoyment of all.

If all these activities were not enough, he had established himself as a builder specializing in military fortifications, notably at Traublay, Rouvray and Toury, so he had a good grounding in the logistics, organization and practicalities of a major building project.[6] Of course, he employed a Master Mason for his work at St-Denis, but he certainly maintained a very personal involvement. Most building projects take the name of the man who commissioned them, giving the impression that that person was actively involved. In Suger's case, this was true. Happily for us, he wrote an account afterwards at the request of his monks. From this we know that he had always wanted to have a go at transforming the church and that the original building was so cramped that on major feast days 'the narrowness of the place forced women to run to the altar on the heads of men as on a pavement with great anguish and confusion' (most possibly an exaggeration!).[7] Work began at the west end of the church and Suger himself wrote the verses that were engraved in copper-gilt on the doors:

For the glory of the church which nurtured and raised him,
Suger strove....
Bright is the noble work but, being nobly bright, the work
Should brighten the minds, allowing them to travel through the lights
To the true light, where Christ is the true door ...[8]

This seems to epitomize Suger's dream. He filled his church with luxurious ornaments and designed the choir in the Gothic style with a vaulted roof such as had been seen at Durham four decades earlier. He was absorbed by light and how that might have a religious interpretation or implication. At the most obvious level, he had the windows painted with biblical stories and recorded their details, even down to which separate stories were shown in the same window. The glass was quite simply marvellous to the people who saw it. They were used to rich, shiny things in their churches, but had never before beheld anything that glowed according to the time of day or the strength of the light outside. It must have been as if God Himself were controlling the messages within them and this will have made an enormous impression on visitors to the church. We are used to seeing light changing through coloured glass; we even use colour to control our traffic and we think nothing of changing the strength of a dimmer switch to get a different ambience. It is not miraculous to us. This new 'nobly bright' work was regarded as being there to brighten minds. The colours were brilliant in a new way because they were translucent, the windows seeming to expand as the sun brightened and so were almost bursting with the religious fervour their stories were meant to convey. This was wonderful, tremendous and terrifying. What Suger might not have expected was the effect the coloured glass would have on the floor. We have all seen light shafting through stained glass and making coloured pools on church floors or walls. What we may not have thought about are its implications; it is reflected light, of course, but in this case, it is light that has been filtered through religious stories and comes from the direction of Heaven. How big a leap of imagination would it have taken for people like Suger to have wondered if this new light within the house of God was not actually carrying the essence of that very God? If one were a true believer, as Suger undoubtedly was, then to stand in such light must have been both awe-inspiring and terrible.

The new choir containing these wonderful windows and other new Gothic features was consecrated on 11 June 1144, and was attended by all the great and good of the land: Louis VII, his wife, Eleanor of Aquitaine (who would later marry Henry II of England), the archbishops of Rouen, Rheims and Canterbury, thirteen bishops including those of Chartres, Evreux, Châlons, Meaux, Sens and Auxerre as well as leading religious figures such as Bernard of Clairvaux. The occasion was so splendid and the new architecture and glass so mind-bogglingly gorgeous that it did not take long for it to become the 'must-have' style for any new cathedral. In fact, as already indicated, what became the Gothic style was already finding favour in other parts of Europe.

Renewed spirituality during a time of change

In our day ideas spread rapidly; we are increasingly led to believe that we can have anything we want as soon as we want it. That was not the case in the Middle Ages. The Romanesque/Norman style was relatively new in Britain, being less than a century old. However, on the Continent it had endured considerably longer. It was not the habit of the Western Christian Church or, indeed, society to swing wildly from one new fashion to another, which makes us wonder whether there were other forces at work and if Suger's architectural experiment just happened to coincide, and, yes, chime, with changes in thinking. If we wonder whether a renewed spirituality was linked to the new architecture, we would have to ask ourselves why there might have been the need for one. It is true that there were changes afoot, that the twelfth century was one of persecution, which always indicates a lack of confidence, that monastic houses were flourishing and new monastic orders arising. The man in the street increasingly called upon the Virgin Mary for her help and religious thinking seemed to be focused more on the humanity of Christ. Yet the Vatican must have seemed to be losing its grip with repeated challenges from anti-popes (those who set themselves up as pope in opposition to the legally elected one for a variety of reasons, but mainly for power, money and thinking that they were better suited to the position, even if they had not been elected). There have been twenty-five anti-popes, ten of whom operated in the twelfth century.

The events of the twelfth century were so varied and important in terms of church administration that they will have had an effect separately and jointly, and the sum of them must have influenced spiritual thinking of the day. A key factor in this would be the state of the Church in Europe. In 1074, Pope Gregory VII had reasserted the Church's authority by banning lay investiture and that had brought him into direct conflict with Henry IV of Germany. The problem was that kings and emperors had assumed responsibility for filling vacancies when a bishop or abbot died, and they often appointed a member of their family or a friend or someone else in exchange for cash. This meant that those in lucrative episcopal positions had often had no theological training and even less interest in doing anything other than collecting revenues. Homage was required of these appointees before they were consecrated, giving a clear message as to where their allegiance was to lie. This applied in England, too, with matters coming to a head in 1099 when Anselm, Archbishop of Canterbury, refused to do homage to Henry I or to consecrate bishops who had received lay investiture.[9] The lack of confidence in the bishops spread down the chain to parish priests who were not all thought to have their minds fully on the job; indeed, many were guilty of immoral behaviour. A major cause of concern at the time was whether or not their work was valid: in effect, if they had been ordained by a bishop who had not been consecrated properly, were they real priests themselves? In addition, if they were not proficient, or were full of sin themselves, was it legitimate to make Confession to them or even to receive Communion from them? Worse than that, if they baptized your child, was he really baptized and if such a priest administered the Last Rites to you, would it be enough to help you get into Paradise? Had you been able to atone or not? These were crucial questions at the time and caused much angst as people were genuinely frightened that they might be permanently barred from Paradise simply because of disagreement and lax behaviour among those in authority. Much depended on the local bishop because it became clear that consistency of religious practice was disappearing. It is noteworthy that up until the twelfth century there had also been a remarkable uniformity in church art right across Europe and Britain. Such art that survives shows clearly that, for example, the march of strange bestial heads on the outside of a twelfth-century church was uncannily similar in style to others such as at Barfrestone in Kent, Kilpeck in Herefordshire, Castor in Cambridgeshire, Studland in Dorset, Poitiers and Aulnay in France, and Santiago de Compostela and Frómista in Spain. This will be considered more closely in Chapter Seven.

The response of the Catholic Church

The Church's response to what must have been viewed as chaos within its walls was to lay down some rules, which were thrashed out at the Lateran Councils, a series of four councils held at the Lateran Palace in Rome in 1123, 1139, 1179 and 1215, each under a different pope. They gradually worked out a framework of administration and a code of behaviour, even going into such detail as how to deal with drunken clerics and forbidding clergy from engaging in secular pursuits such as playing games of chance, visiting taverns or attending unbecoming exhibitions: priests were expected to live virtuously. Matters must have got into a bad state if they had to consider that level of legislation! An annual conference was to be held to check any abuses within churches, not just in the sense of following the correct liturgy, but also mundane things such as storing household goods in the church building. Some individuals such as Abélard, whose writings had been publicly burnt at Soissons in 1121, challenged the Church. Amongst other things, Abélard had made a collection of quotations from the Bible and from the writings of the Fathers of the Church that seemed to him to show inconsistencies in teaching by the Christian Church. He arranged his findings in a compilation called *Sic et Non* ('Yes and No'). He explored every possible theological argument and managed to cause the utmost offence to large sections of society and the Church along the way.

Monastic life was attractive in the Middle Ages. It might not appeal so much in the twenty-first, but then, if someone had no family or support group and a strong religious belief (or even if they did not), it must have seemed a sensible solution. Holy men and women were held in high esteem and monasteries were associated not only with spirituality, but with status. The twelfth century, there-fore, brought not only Gothic architecture and an increase in challenges, but also a plethora of new monastic orders. The last two items seem to contradict each other, but they also show a breaking away from the secular Church in dif-ferent ways. As a reaction to what must have seemed to have been a maelstrom of dissidence, impiety and dissolute learning, the Church did two things almost as though it were trying to attract the attention of lay people. First, the Church instituted new feast days and, second, it encouraged the cult of the Blessed Virgin; never before had Christian worship in Europe been so Mary-orientated.

In Christianity, Mary is the mother of the Son of God and, therefore, extremely powerful in her own right. To this day, St Mary is the most popular church dedication in Britain, outnumbering the next most popular, All Saints, by two to one. She represents the human face of God; someone who would under-stand the tedium of daily life and its sorrows because she had lived them. She had been one of us, but also *not* one of us since she had been so pure that she had been chosen to carry the Son of God. Perhaps one of the most important things about her was that she had lost a son, something all too many people could relate to in the Middle Ages. She is often shown as a tender mother, never more so than on a thirteenth-century *voussoir* of one of the west doors of Rheims cathedral, where she looks surprisingly relaxed and modern as she swaddles the Holy Child on a wicker basket. At the end of her life, she was taken up into Heaven on the Day of Assumption (15 August), where she was crowned Queen of Heaven by her son. The story of the Assumption had already become popular in the tenth century – indeed, the first depiction of it in Western art is in the Benedictional of St Æthelwold made at Winchester after AD 971.[10] This was much assisted by a nun called Elizabeth of Schönau, who had a vision about this very subject in the late 1150s, just when the cult of Mary was getting into its stride.

At this time, rosaries came into vogue. They are said to have been invented by St Dominic *c.*1200 in Spain and, although their origin is not clear, they were certainly used by Dominicans and Cistercians alike to promote devotion to Mary. Likewise, the Angelus became popular, although it is only possible to trace it back to thirteenth-century Germany with any certainty. This prayer, popularly called the Hail Mary, became a central part of Catholic worship and remains so to this day. It sums up the continual fear felt by our medieval ancestors that no one would be on their side when they left this world: 'Pray for us sinners, now and at the hour of our death.' Mary became a crucial part of medieval spirituality and yet at times believers seem to have treated her somewhat casually. The particular carving of her at Rheims is a tiny part of the central west door that has to be hunted for among scores of other images, some of which are not even religious, which supports the idea that Mary was seen as one of the people. It is plain that no disrespect was intended because she also features in the main image on the west façade, which is one of the most graceful crownings of the Queen of Heaven in Western art, and the pointed arches of Gothic architecture lend themselves perfectly to her story.

Not all of these events would have been directly connected to the new architecture, but they provided the conditions under which it could flourish. In the twelfth century, spiritual matters both fragmented and gelled. At the start of the century there was one type of church architecture mainly in use across Europe. By the end of the century, church architecture had changed utterly. At the start of the century, very few would have thought of challenging the Church as individuals and groups in the way that Abélard and others would. Universities and some monastic orders were in their infancy. By the end of the century there had been a rapid change in church decoration, which has mainly been laid at the feet of Abbot Suger. It was all very confusing to the ordinary man and woman, whose only real aim was to get through this life and guarantee a place in the next. As far as the Master Masons and their employees were concerned, it meant that there was a lot of building work to be had.

CHAPTER TWO

PAVING THE WAY

WHEN WE LOOK at a cathedral, we see a building that has stood for centuries and which is unlikely to change dramatically in the future unless parts of it fall down. Given constant survey, monitoring and restoration projects, let us hope that this will never happen. To us, therefore, a cathedral is the finished product. New glass, statues and memorials will be added, of course; short-sighted birds will continue to wreak havoc on unprotected glass; and, as York and Peterborough know only too well, there is always the danger of lightning and fire. However, the days of major planned change, of adding whole new wings or chapels, are probably over. It is easy to forget that what we see can be rather different from what was originally envisaged, although we can track changes through the varying styles used. Happily for us, successive architects were keen to use the latest developments and techniques so we can see a progression from, say, Romanesque/Norman to Gothic in one building in almost every cathedral that dates to the pre-Gothic period. Sometimes this is dramatic, such as at Vézelay where the nave is solidly Romanesque, but the open-plan choir is all slender pointed arches and light. More often the styles blend effortlessly as at Peterborough, Norwich, Durham, Winchester, St Albans, and numerous others. It would be interesting to know what the medieval man in the street thought about additions to cathedrals. Certainly, many people today find that a very modern building too close to an ancient cathedral jars on the senses, although is it surprising how frequently an ancient building sets off an ultra-modern sculpture or stained glass to its best effect.

We do not always know how the terrain looked before building began, but we do know that it was often land on which houses and businesses had already been built and which had to be cleared. In twelfth-century Durham, Bishop Flambard had the area in front of the cathedral forcibly cleared so that the church should be 'neither endangered by fire nor polluted by filth' when he wanted to make the space now known as Palace Green. Most cathedrals were built on the site of a previous church but take up much more room; they are bigger and grander than the first church would have been, just as many of those churches were built on pagan sites in the first place, thereby claiming that ground as Christian. It is almost mind-boggling to imagine not only starting to build something of that size, but also embarking on a project that was unlikely to be finished within the Master Masons' lifetimes, never mind working out the costs and logistics of it all. But this is something that they did – indeed, still do, as will be discussed later. However, now we should think about the building site itself and how any cathedral came about in physical terms.

Choosing a Master Mason

When planning a building today, one of the first things we would tackle would be the ground on which it was to be built and the actual building design; indeed, the site chosen will have a bearing on the building design. A committee or council deciding to build a new hall, for example, would have a clear idea about what would be required in terms of function. In cathedral building that is even easier because the primary function is obvious; therefore, the building must have a main altar, choir, nave, etc. The initiative might come from a group or from an individual, but everyone would want the best available. Bishop Guillaume de Seignelay of Auxerre, in France, wanted a new church in 1215 that 'would not be inferior to those others in form and treatment'. A fundamental difference at the planning stage is the way in which Master Masons presented their plans to the Workshop Committee. The Workshop Committee would be set up early in the proceedings to be the focal point for all direction, enquiry, supervision, allocation of funds and inspection of works. This committee would be separate from the cathedral's usual administrative body (known as the Chapter), although it would include at least one representative from the Chapter. It is occasionally thought that the committee was made up entirely of clergy from the cathedral, but both the Church and the builders were more professional than that and found the Workshop Committee's members from people who had experience and understanding of both building and financial administration. They were, after all, about to spend colossal sums of money and would not simply trust these matters to luck. By the time Gothic architecture was in vogue, the world had moved on; professionalism mattered. Just as lay scribes and artists were regularly employed to illuminate manuscripts rather than allowing such activity to be the sole domain of scribal monks, so it was no longer unexpected that an enterprising monk or bishop should add building to his list of skills, although a few did.

The man in charge of the Workshop Committee was variously called the Master of the Works or Master of the Workshop, which can cause confusion because the Master of the Works was *not* the Master Mason. Indeed, the Master of the Works was the man charged to find the Master Mason, who would then run the project on the ground and turn ideas into realities.[1] At this point, it is hard to get an idea of the timeframes involved and plainly these varied. In the twenty-first century we can find an architect's office quickly and perhaps speak to one the same day. In medieval times, the Master of the Workshop had to invite Master Masons to submit their ideas so, if the plan was for a relatively small church, he might be content to go to a man with a sound reputation who lived locally. For a cathedral or large church that was to be internationally admired, of the sort that would attract pilgrims and visitors and would reflect the status of the city, a different level of expertise and experience would be required. Some Master Masons were invited to compete for the job while other candidates will have arrived to try their luck. The Master Mason did not necessarily report to the Committee with ready drawn-up plans, but instead with the ability to persuade and a model of the type of building he envisaged. Some of these survive, such as the one for a church at Regensburg. No doubt the models would have been made well in advance as part of a Master Mason's marketing material and interview kit; nonetheless, he might have travelled hundreds of miles from another country to see if he could be appointed. After the disastrous fire at Canterbury in 1174, the damage was so great that we are told that people were maddened with excess of grief and perplexity. Gervase, a monk at Canterbury, who left a graphic account says, tellingly:

> However, amongst the other workmen there had come a certain William of Sens, a man active and ready, and as a workman most skilful in both wood and stone. Him, therefore, they retained, on account of his lively genius and good reputation, and dismissed the others.[2]

How many others applied is not known.

Gervase also tells us that this French mason spent a long time considering the task after he had been appointed, making his own plans in private and possibly persuading the monks to try out new ideas. Evidence for this can still be found in Canterbury cathedral itself: on the south wall are carved a row of three blind arches, which serves no decorative purpose. It is thought that these are templates made by William of Sens to persuade the monks that they should go for the new-look Gothic style since two are of a slightly clumsy, round-arched, Romanesque/Norman design and the other is more elegantly pointed. The cathedral at Sens, built between 1140 and 1168, can claim to be the first ever whole Gothic cathedral since it is the first one made to the new style in entirety, as opposed to having a reworked choir or other part of it. Thus, William of Sens' ideas will have been as up to date as possible, and he could claim to have a sound architectural standing.

In general, though, the Master-Mason-designate reported to the Workshop Committee with a model of the general style of building he would make, but he also had to convince them that he was the sort of man who could coordinate the whole project and see it through, which was no small undertaking. He also had to agree not to change any plan that had already been agreed. The statutes of Saint Michael of 1563 were very clear on this:

> If someone signs a contract for a project and presents a plan for the way in which it is to be executed, the work may not in any way be modified in relation to the original design. He must execute it strictly according to the plan that was shown to the patron, whether he be a lord, a city or an individual, so that nothing is missing from the building ...[3]

Finding a building style to suit

Actual drawings for parts of the building were produced as the project progressed, and some of these show minute details, such as the one for the façade of Strasbourg cathedral, built around 1360. Sadly, very few of these drawings survive in England. The Academy of Fine Arts in Vienna has a remarkable collection of 425 Gothic architectural drawings, but these are rare, not least because they were not always drawn on parchment, but on walls or even on the floor. References to drawings occasionally surface. A carpenter called John Berewik was contracted to make a frame for a timbered house for the Warden of Winchester College in the mid-fifteenth century. The house was 'to be as made as the tracing schoweth as drawn in a parchment skin' and a Master Mason called Henry Pays bought 'four sheep skins for patterns' in 1473 in Suffolk.[4] In the accounts for the abbey of Vale Royal in Cheshire, a payment was made in 1278 to workers who levelled and smoothed the ground so that a plan of the monastery could be drawn, and this was likely to have been full size, so the patron must have taken a bit of convincing about the sizes of rooms, and so on. This is very human of him; many of us have paced out spaces in our own homes to see if items will fit. Above the Chapter House in York Minster there is a room known as the Mason's Loft, which has a plaster-of-Paris floor on which plans were marked out so that the Master Mason could brief his workers – and this is not the only example that has survived from this time. When that stage of the building had been completed, the floor could be resurfaced and used again like a gigantic blackboard. Luckily for us, one Villard de Honnecourt (of whom more in Chapter 9) made numerous drawings of what he saw while Rheims cathedral was being built in the thirteenth century. The same principles are used to this day. On the night of 12 January 2017 Storm Egon swept across Europe, claiming at least three lives. The immense winds crashed through the rose window on the west façade of Soissons cathedral, destroying centuries-old glass and shattering the tracery. As it happened at 10 o'clock at night, no one was hurt, but what stones could be retrieved were laid out on the cathedral floor on top of a real-size tracing (*see* plate 4). In due course, new stones will be cut to fit in with the old and the window restored to its former glory, no doubt at immense cost. It can hardly be a consolation to the custodians of the cathedral to know that it is fascinating to see medieval techniques at work to this day.

Despite the strictures mentioned above, it is strange to realize that the initial plan was often only an indication of what would finally be built. All cathedrals have been much added to over the centuries, but the first Master Mason does not appear to have been required to come up with all the solutions to complete the building at the outset. Bizarrely, it seems that competitions were held when a problem was encountered, most famously at Florence, where Filippo Brunelleschi won the contest with his design to cover the huge central space of the *duomo* with a domed roof, but one wonders how the building could have got as far as it did without that problem being addressed. Supposing that no solution had been found! Competitions were also held for less important details, such as how to decorate capitals on the tops of columns and even what would be their final height. This had the advantage of allowing new talent to be discovered to the benefit of the building, haphazard though it was. The Chapter and Workshop Committee were keen to be as up to date and showy as possible, as is seen by the phrasing of the bishop of Auxerre's remarks above about his cathedral not being inferior to others. It may seem a perilous way of going about things, but we should also remember how long these projects took; Chartres was one of the fastest builds, being more-or-less completed in a quarter of a century. It could be argued that many have never been finished at all. It is thought that Rheims cathedral was originally meant to have had seven spires giving it the fairy-tale look achieved at Laon when seen from a distance. The spires were never built either through lack of will or money, or anxiety about structure. The truncated-tower look seen on so many of our cathedrals became the norm and, subsequently, the fashion in itself. Salisbury cathedral, with its lovely single spire, could be said to be one of the few that is actually finished as it was only ever intended to have one spire.

Precise measurements were less easily made, but a great interest was taken in geometry in the Middle Ages. Along with arithmetic, astronomy and music, it was part of the quadrivium, in turn part of the Liberal Arts that formed what we would term 'higher education'. There was an almost mystical belief in numbers and their symbolism and what might be learned from their correct use and understanding. Perfection and visual harmony might be found if the proportions were right, so many cathedrals' ground plans are based on a series of squares and triangles that link to each other. In many cathedrals a line drawn diagonally across the square of the cloisters equals the length of the nave. If the nave is made into a square and a line is drawn diagonally across that, it might be equal to the length of the whole building. To a similar end, the so-called 'golden' section, a traditional proportion which was supposed to express the secret of visual harmony, was often used, especially in Italy.[5]

Preparing the ground

Having got as far as finding a style that suited all concerned and a Master Mason capable of producing it, two things had to happen: one was to find materials and workers; and the second was to prepare the site. Given that the cathedrals we see are built in the very centre of cities, it must not be presumed that there was always a convenient space for them (and the town will have expanded around them). They may well have been built on top of an old church, but these new buildings were gigantic and not only needed space for themselves, but probably also for a new bishop's palace next door and/or administrative buildings. Most certainly they needed space for the builders to work in ranging from accommodation huts to workshops and perhaps even to quite substantial ponds to provide water for the mortar mixers. Although a cathedral brought great prestige and certainly business, it should not be thought that everyone welcomed work going ahead.

There are accounts of houses and barns being demolished to make space and of fields being beaten down to make roads, which inevitably must have meant that people's livelihoods were damaged. At Winchcombe, Gloucestershire, in 1245, this was managed in a civilized manner with a public inquiry as to whether:

> it would be to the injury of the town or of the Abbey of Winchcombe, or of anyone else, if the King should allow Master Henry, rector of the church of St Peter of the same town, to lengthen the church twelve feet eastwards, and to enlarge an aisle begun on the south side of the church to the length of thirty feet, and the breadth of twelve feet ...[6]

Elsewhere, things were handled less democratically. In 1094, the first bishop of
Norwich, Herbert Losinga, who evidently was not without funds, simply:

> bought for a great sum a large part of the town of Norwich and, having torn
> down houses and levelled the ground for a great space, built in an excellent
> position on the river a most beautiful church in honour of the most high
> Trinity.[7]

Even before the institution of modern planning permission, the medieval Church
could not usually ride quite so roughshod over others: Losinga had to pay; he
was not able just to take. In some cases, perhaps the local lord could simply take
what he wanted, as suggested by an account of Arnoul II, count of Guines, who
had decided to fortify Ardres. It appears that Arnoul even cut down flowering
trees and broke up gardens and flax. 'He did not bother about those who became
angry and cried out.'[8]

If you had a business in the centre of a city close to the original main church,
you were probably prosperous so you could afford lawyers to protect your
interests, and you probably had the will to do so. The poor folk sheltering by
the walls of the original church building just got moved on. The Workshop
Committee would have included compensation in their initial budget and no
doubt pleas would have been lodged by surrounding landowners in advance
once they realized that roads might be made across their land. That said, they
may not have minded too much since they could easily make good any losses in
the long run simply by imposing tolls to use those roads. The ground recovered
in this way would then have to be prepared, either in the removal of buildings
and rubble or by being rammed down if it was at all marshy. At Ramsey, in the
tenth century, they began by beating the ground as hard as they could 'with
frequent blows of rams'.[9] This failed and they had to start from scratch with
proper foundations.

Obtaining the right materials to build

Next, a Master Mason would need a quarry and a forest, ideally as close to the
building site as possible for reasons of cost. The amount of stone needed is beyond
easy imagination, especially when one takes into account the number of churches
and cathedrals being built; it is no exaggeration to say that the centre of all major
European cities would have resembled building sites for decades at some point in
the twelfth to fifteenth centuries. As Jean Gimpel has remarked:

Within three centuries, from 1050–1350, several million tons of stone were quarried in France in order to construct eighty cathedrals, 500 large churches, and several tens of thousands of parish churches. More stone was cut in three centuries in France than in any period in the history of Egypt, even though the Great Pyramid alone has a volume of two and a half million cubic metres.[10]

As early as the eleventh century there was so much building work taking place that the monk Raoul Glaber observed that:

It befell almost throughout the world, but especially in Italy and Gaul, that the fabrics of churches were rebuilt, although many of these were still seemly and needed no such care; but every nation of Christendom rivalled with the other, which should worship in the seemliest buildings. So it was as though the very world had shaken herself and cast off her old age, and were clothing herself everywhere in a white garment of churches.[11]

One can see clearly what he meant by visiting a cathedral that has recently been cleaned on the outside: the white stone of Auxerre was almost blinding in the summer sun of 2008 and must have been even more dazzling to medieval eyes, with its marvellous carvings and height and promise of salvation. It still makes a very attractive picture above the river Yonne but, when it was first built, the sight must have been astonishing to locals and visitors alike, and must have confirmed the existence of Paradise to the pilgrim.

Merely finding the quarry took time, and money would be allocated for the search. It was more than just finding a quarry; other considerations, such as the quality of the stone available and the crucial question of whether or not it could produce enough to finish the job, had to be considered. The last thing a Master Mason wanted was to have to change to a different stone, making the cathedral look like some sort of badly baked cake. Specifically, we see a man called Hugh de Hedon visiting quarries at Thevesdale and Stapleton in 1399 to source stone for York Minster.[12] In 1413, Richard Winchcumbe was paid to go to Taynton to check stone for building at Adderbury and, nearly thirty years later, William Hobbys was paid six pence per day for eight days to visit quarries at Upton and Freme, 'to choose and examine good stones called Cropston' for the repairs at Gloucester Castle.[13] Time spent in this valuable research would always pay off; no one was going to run the risk of having to finish the building using something different, making a mockery of the overall effect and ruining their reputation as a designer. That it was the responsibility of the Master Mason is made clear in William Vertue's contract for work on St George's Chapel, Windsor in 1511:

Also the said William covenants and grants by these presents to find all
manner of stuff as stone, lime, sand, timber, scaffolds, boards, nails and all
other things necessary for the same work with carriage by water and by land
and all manner of workmanship necessary and appertaining to the same.[14]

If a Master Mason was lucky, he might have a quarry nearby, such as at Barnack,
which provided stone for churches in Lincolnshire. Stone coming from the
quarry at Purbeck was more specialized and tended to be used for decorative
purposes. It is evident that various masons became adept at working with different types of stone in diverse ways. To own a quarry as the Master Mason,
Robert le Hore, did at Bentley in the fourteenth century must have been similar
to owning an oilfield. Once the building was started, the Master Mason had
tremendous bargaining power, and medieval businessmen became adept at
drawing up contracts from the outset of the project. The Chapter and Workshop
Committee of Laon were lucky in that in the twelfth century they were able to
buy the quarry at Chermizy some six miles to the south-east of Laon. Definitely
the best place to quarry was considered to be Caen in Normandy, which provided stone for numerous cathedrals in France and some in England, most
notably Canterbury, although there may have been some victor's nostalgia in
that choice. It was not, however, a ridiculous choice since the cheapest and most
efficient way to transport stone was by water rather than by road; stone could be
taken across the English Channel and then by river to Fordham, a short distance
from Canterbury. That said, keeping costs down was important. When William
the Conqueror decided to build his votive abbey at Battle, near Hastings, the
stone was originally to be brought from his native Normandy. However, he was
not so blinded by sentiment or desire to impose Norman materials on England
that he felt able to ignore a local woman who reputedly had a miraculous dream
that showed where suitable stone could be found nearby. It would be cynical to
suggest that her dream may have been more to do with the local economy and
collaboration, than with anything more spiritual.

Elsewhere we hear of abbots praying to have such guidance granted to them,
which shows what a godsend (perhaps literally) a local quarry could be. For
example, at Cambrai in the eleventh century where Bishop Gérard, their bishop,
was vexed by having to bring stone that was suitable for columns from some
thirty miles away until divine intervention showed him a place that was considerably nearer. At Exeter, they were able to use stone from the quarry at Beer,
and this was also taken further afield to Westminster and Winchester. Beer stone
was popular because it has a good cream colour and is soft so it is easy to carve,
but it hardens on exposure to the elements.[15] Having secured his stone, a Master
Mason would then have to safeguard it. Legal rows broke out in 1251 when stone

that had been collected at Bristol for the abbey was seized by the king's men for use in a castle. The monks won the day and got their stone, but this added to the expense and held up work considerably. Henry III put an embargo on stone being sold or carried by ship from Kent in 1246 unless it was going to be used on his own building project at Westminster. Henry III was renowned as a great builder. His works stand for us to see, but he caused dismay by what seemed to the people to be his frittering away of money that would have been better used elsewhere. By 1261 he had spent £29,345 19s 8d on the work at the abbey church of Westminster alone, and it is possible that the final sum he spent was in the region of £50,000.[16] It is thought that his interest in architecture was awakened when he married Eleanor of Provence, whose family was known for its patronage of the arts in general. He became such an enthusiast and so knowledgeable that he is sometimes portrayed holding a model of one of his buildings; for example, on the Gallery of Kings on the west front of Lichfield Cathedral.

Similarly, wood was an issue and must have led to considerable deforestation in areas where daily usage alone was already having a major impact. In the absence of any other source of energy, wood was needed for anything at all to do with heating and cooking as well as for building, fencing, making barrels, carts, ploughs, tools, domestic utensils, furniture, military equipment, mine-props, and any industry that needed heat such as a forge. The timber industry employed huge numbers of people ranging from foresters and wood-cutters, to tree-splitters and boarders, and all the various trades concerned with transport-ing it and looking after the livestock involved. When it came to the scaffolding and structure of cathedrals and other large-scale buildings, enormous areas of forest were destroyed. Almost four thousand trees were used at Windsor Castle in the fourteenth century.[17] Wood was needed for numerous aspects of building, not least boards for temporary and permanent floors, ceilings, panelling, and so on. These could be bought ready-made, but obviously still involved the finding and cutting down of good-sized trees in the first place. One thousand boards were used to make a lodge at Vale Royal Abbey, so this was not only big business but another major demand placed on forest resources. Trees – mainly oak – of considerable maturity were required for central beams, as can still be seen if one climbs up into the roof of Ely or Salisbury cathedral. In 1140 Abbot Suger was looking for such trees for the abbey of St-Denis to the north-west of Paris and wrote an account in his *De Consecratione*:

In order to find the beams, we consulted all those who work with wood in our area as well as in Paris, and they all replied that … because of the lack of forests, beams could not be found in these regions, but we should have to get them from the region of Auxerre … We were overwhelmed at the prospect of all the effort and the great delay that this would involve. One night, having returned from matins [night office], I thought as I lay on my bed that I should myself go around the neighbouring woods … I set off early in the morning with the carpenters and the dimensions of the beams and made my way quickly to the forest of Yveline. Passing through our lands in the valley of the Chevreuse, I called together … those who were in charge of the land and all those who knew the forests well, and I asked them under oath whether we had any chance of finding beams of these dimensions in that area. They began to smile … astonished that we should not know that nothing of that size was to be found in all those lands … We ourselves rejected everything that these men said, and with bold confidence we set out through the forest; in about an hour we found one beam of the right dimensions … At about nones [office said at 3pm] or a little earlier, pushing our way through the woodland … through the thorn bushes, to the astonishment of all those present, … we discovered twelve beams. It was the number that we needed.[18]

Well might Abbot Suger have felt dismay initially since the distance from Paris to Auxerre as the crow flies is one hundred miles and he would already have taken into account other materials needed, which would have needed transportation over great distances. Some cities specialized in producing items such as cobalt oxide for glass or lime for mortar and if iron was needed, that generally came from Spain. The Workshop Committee was not above self-help and, if possible, would recycle material, such as the city walls at Beauvais, which were reinvented as part of the cathedral.

Finding a willing workforce

Self-help was a recurring theme on medieval building sites. Bishop Hugh of Lincoln and Louis IX (later Saint Louis) of France were both said to have got their hands dirty by pitching in to help with labouring, so great was their desire to be part of making the house of God. It seems likely that the reality was more symbolic than actual, especially in the case of Louis IX, spiritual man though he was. These stories circulating all helped to encourage others to do likewise so that at Le Mans great numbers of people rolled up their sleeves when the body of St Julian was to be translated to a new choir in 1254. Work had evidently just finished in time for the ceremony to take place because the description by the bishop, Geoffrey of Loudon, talks about needing to free the church of building debris. We are told that people of every condition, age and sex were there, carrying out rubbish and were:

Vying with one another in their eagerness. Matrons were there with other women who ... not sparing their good clothes, carried the gravel outside the church in various garments, in clothes bright with green stripes or some other colour ... others, filling the tiny garments of babies with the rubbish from the church, carried it outside the church. Small wonder. It was fitting that the praise of infants attend the divine work. In order that infants and little children might seem to have contributed their labour to so great a service, three-year-olds and little children in whom one could already discern signs of holy faith and who until then could scarce walk carried the dirt outside the church in their own little garments. Those who were older carried great pieces of wood and stone outside the church more quickly and easily than could be believed ... In a short time they did voluntarily what many hired men had not accomplished over long periods of time. And this they did on their holiday ...[19]

One hesitates to suggest that Bishop Geoffrey had a rose-tinted view of what actually happened and it surely must have been more useful to use sacks than baby clothes to carry debris, but it is a charming story and not untypical. Elsewhere at the abbey of Saint-Pierre-sure-Dives, Brother Haimon asked:

Whoever saw, whoever heard, in all the generations past, that kings, princes, mighty men of this world, puffed up with honours and riches, men and women of noble birth, should bind bridles upon their proud and swollen necks and submit them to wagons which, after the fashion of brute beasts, they dragged with their loads of wine, corn, oil, lime, stones, beams and other things, necessary to sustain life or to build churches, even to Christ's abode? Sometimes a thousand men and women are bound in the traces (so vast indeed is the mass, so great is the engine, and so heavy the load laid upon it), yet they go forward in such silence that no voice, no murmur is heard ... no other voice is heard but confession of guilt, with supplication and pure prayer to God that He may vouchsafe pardon for their sins.[20]

He then describes how all quarrels, grudges and debts were set aside in this mass activity. Spontaneous assistance is a miracle-style story that goes back into the history of many countries. At Lindisfarne in 1093, St Cuthbert inspired such devotion that local people dragged loads of stone over wet sands to help the monk Edward to build a new church in his honour on the island. Prior to this, angels might be employed as labourers; Bede tells us how St Cuthbert himself was helped by angels to move oversized stones on Farne Island when he became a hermit there in the seventh century, a story reiterated in the building of the monastery at Mont-St-Michel. Some of these cooperative efforts appear to have been highly staged and were none the worse for that; many later ones were given impetus by a specific event, such as at Chartres in 1194 when a huge fire on 10 June all but destroyed the city and the cathedral – only the crypt, the two towers and the Royal Portal survived. Previously, there had been trouble in the city because some people were exempt taxes while others had to pay, and this trouble had resulted in brawling in the street. To try to calm things down, the pope had sent a legate to arbitrate, among whom was Cardinal Melior of Pisa, and it was during this visit that the fire occurred.

What was particularly hard for the people of Chartres to take was that, in addition to the loss of their family and friends and their homes and businesses being destroyed, their famous relic, the veil that Mary wore at the birth of Christ, was in the crypt and presumably reduced to ashes. Given the religious attitudes of the day, not to mention the horrible ordeal they had just endured, the people were deeply affected by this and took it to be another judgement upon them for their sinful lives. Cardinal Melior was a great orator and he was also a superb stage manager. He brought the people together among the smouldering ruins of their once beautiful city, ostensibly to console them over the loss of the relic and to put heart into them, but unbeknown to the general public, the relic had not been lost. When the fire had broken out, two monks had rushed into the crypt and bolted the metal doors. It is best not to try to imagine what that night must have been like for them with the heat, the smoke and the disagreeable sensation of the cathedral collapsing above them. It was believed that they were dead and justifiably so, but, remarkably, they had survived, the undisputed heroes of the hour. In an outstanding piece of theatre, at the very moment that Melior was reaching the climax of his speech, the Chapter of the cathedral came into the midst of the people bearing the relic aloft. It did not then take much for Melior to persuade the people that this was a sign from the Virgin Mary saying that she wanted them to build her a new shrine.[21] It was also Melior who persuaded the bishop and Chapter to give a large part of their personal salaries to the rebuilding for a period of three years, further proving his remarkable oratory skills.

There was huge rejoicing and an immediate general agreement that the Virgin Mary must want Chartres to be her spiritual home, and that therefore the biggest, brightest and most beautiful cathedral ever seen must be built here. So spontaneous an agreement was this that people voluntarily went to the quarries five miles away and dragged the carts laden with stone themselves. Not only was this deemed to be one of the Miracles of Mary that were written about and circulated almost immediately, but it also triggered other spontaneous assistance stories throughout the twelfth and thirteenth centuries. How long these bouts of assistance lasted is not recorded and, human nature being what it is, it seems unlikely to have lasted more than a day in reality with perhaps a handful of people working longer. It is interesting to note that over the centuries there is a shift in surviving records from chroniclers recounting miraculous intervention to straightforward contracts being drawn up to avoid having to rely on such a contingency.

Transporting materials to the building site

Whatever their input, any volunteers were invaluable because the costs and difficulties of transportation and procurement involved were phenomenal. It has already been mentioned that the easiest and quickest way to move stone was by water. Even though it meant loading and unloading barges, if possible it was best to travel by river and, indeed, at Bury St Edmunds and elsewhere a canal was built for exactly this purpose. Oxen were used a great deal, being the strongest beasts available and much helped by improvements in harness. Teams of oxen were a familiar sight, dragging carts laden with stone sometimes over great distances. Horses were also much in demand once the shoulder collar was invented, which allowed a team of horses to pull as much as two and a half tons in weight. They were quicker than oxen, although the latter could haul a greater load. One pair of oxen could cart around 1500kg of stone in a round trip of ten miles in a day. This is not a lot of stone for a cathedral; it would perhaps not even make a whole pillar, which gives an idea of just how labour-intensive the building was in terms of man and animal days. It also added enormously to the cost since the drivers had to be paid and the animals fed and secured. There is a record of twenty-six pairs of oxen pulling one cart laden with capitals, which had been cut at the quarry. This may seem huge and hard to credit, but we know about it because it ran over the Master Mason of Conques and broke his legs, so he certainly believed it. An ox-train like that would have been virtually unstoppable once it got into its stride and would need a lot of room to manoeuvre. The enchanting outline of Laon is enhanced through being on top of a steep hill. Their quarry was six miles away on the surrounding plains and the Master Mason was so impressed by the way the beasts laboured that he honoured them for all eternity in stone. They are not immediately obvious at first, but once spotted, are never forgotten (*see* plate 5). Sixteen (two teams) of full-size oxen, all in different poses, were made and they gaze out forever from the western towers from the hill up which their living ancestors had toiled.

A way of reducing the load was to have stone cut to size at the quarry. Once the stone had been identified as being suitable, it would be quarried and scappled, i.e. cut roughly to shape. The Master Mason would have templates or moles drawn up that would be sent to the stone-cutters working at the quarry. The stone-cutter cut the stone; the mason designed the shape of it and placed it. No doubt this distinction led to numerous strenuous discussions in taverns as to which craft was the more skilled and important. If the stone had been cut or placed badly, rain could saturate fissures that would then crack in winter frosts and this could be a major cause of damage and crumbling later on, shortening the life of the building. No doubt both professions blamed the other if this happened.

When William of Sens was working on Canterbury cathedral he sent templates back to the quarry at Caen in Normandy to try to reduce the costs of transportation. Two shiploads of Caen stone for Westminster Abbey cost £24 18s in 1252; in 1429 Caen stone for London Bridge was bought at 2s 6d a ton, but its carriage to London cost an additional five shillings, twice the cost of the actual material. When the Bell Harry tower was built at Canterbury in the late fifteenth century, £1,035 was spent on it, of which £388 15s 6½d was for Caen stone with carriage, cartage and customs. Some 1,132 tons of stone was used on this project at various prices. Buying stone in England did not necessarily make it that much cheaper: in 1278 nine shiploads of Corfe stone from Dorset were bought for the Tower of London at a cost of £48 11s 6d. Thirty-six tons of Beer stone for Rochester Castle cost £18 in 1367.[22] The ton (or tonne-tite) was calculated as being £2,240, or the equivalent of the weight of a tun of wine.

The cost of carriage was so great that Caen stone used at Norwich cathedral cost four times what it did at the quarry, whereas the accounts for Troyes cathedral reveal that stone from Tonnerre went up fivefold. This money was spent on oxen and drivers, carters, loading and unloading, which often involved the use of cranes and the ubiquitous road and bridge tolls. If a Master Mason had to use a road crossing someone's land, he had little choice but to pay, and this affected the general public and merchants as well. One of the few groups exempt from tolls was pilgrims, which is why there are so many cases of merchants dressing as pilgrims to keep their own costs down. The Workshop Committee often tried buying off the toll-road owners with promises of eternal bliss, perhaps allowing a memorial in the new church or other inducements, but that does not often seem to have worked. Quarrels and lawsuits broke out when those who owned the land tried to bar passage across it. The abbey at Bury St Edmunds had the right of quarrying at Barnack near Stamford. William the Conqueror himself had to step into a row and forbid the abbot of Peterborough from preventing the Bury officials from carrying their stone to the water for transportation.[23] To try to stop costs from escalating after work began, the Workshop Committee might pay the Master Mason a lump sum out of which he would then pay junior workers. This was to help persuade Master Masons not to have extravagant ideas that had not formed part of the original budget. We see this specifically in John Wode's contract of 1438 in which he agreed to build the western tower of Bury St Edmund's abbey church and undertook to find the wages, food and drink of all the masons employed at the rate of three shillings per man in winter and 3s 4d in the summer.[24]

Wages in the building trade

Wages were a constant financial problem; it is thought that forty per cent of the cost of building a cathedral went on salaries and casual pay. Wages were so variable according to the particular trade or craft, not to mention the man's individual seniority and proficiency within that trade or craft, that it is not helpful to cite only a handful of examples. Following the massive depopulation through plague, the negotiating power was in the hands of the skilled labourer and, to try to keep an even keel, Statutes of Labour were passed in 1351 and 1388 to try to fix wages so that landlords could be punished for paying too high a rate. Nonetheless, there are cases where this was blinked at, presumably in the interests of getting the work done. Some masons were repeatedly fined eight or twelve pence, or even two shillings, such as Richard Sallynge in 1357–9 for having overpaid his workers. Some judges dismissed the cases brought against Master Masons as it was deemed impossible to set down what their true worth should be. Such a case was John Sampson in 1390 and 1391 on the grounds that he was a:

> Master Mason in freestone and capable and skilled in that art and in carving, and because on account of the high discretion and knowledge of that art, the wages of such a mason cannot be assessed in the same way as the wages of masons of another grade and status.[25]

At other times there were complaints that ordinary masons were being over-paid and not producing the goods, but it was ever thus. A man's pay might also include clothing or gifts, or perhaps have an element for the maintenance of his equipment. Tools were provided at some building sites. At Vale Royal Abbey in 1278, twenty-four hatchets were bought for the masons at five pence each, as well as hammers, wedges, picks, hoes, spades and trowels and, in addition, the Workshop Committee had to budget for the sharpening of tools – as many as 3,822 items were paid for between May 1481 and November 1484 at Kirby Muxloe castle. Gifts of clothes in the form of an annual cloak or robe of very high quality would be given to the Master Mason as well as more practical items such as leather gloves for the stone workers or leather aprons, or even straw hats in a heatwave. Beer and banquets of an informal nature might be given to the workers on feast days, all of which cost money.

By the time the cathedral was ready to receive high-quality art, perhaps in the form of altarpieces, then serious money would be spent. In Siena in 1311, Duccio was paid 3,000 gold florins for his magnificent altarpiece, which took him three years to create and was carried in great devotion to the cathedral in a procession. The whole town appears to have turned out with people processing around the Campo, while bells rang joyfully. At Westminster Abbey the costs, less the nave, between 1245 and 1272 came to £40,000. A new wooden lead-covered spire at St Albans in 1297 was £248, but the felling of 620 alder trees at Lichfield's Bishop's Palace in 1308–9 cost only twenty-nine pence.[26]

Financing the construction of a cathedral

With all this expense, then as now, fund-raising was a major issue. There appears to have been a difference in the general patronage of religious buildings according to country. Very broadly, in France only a few kings gave to the church with the notable exception of the saintly King Louis IX, although they might donate a window or statue to commemorate their own life or that of an ancestor. In the area we now call France the Church tended to organize its own building. In England, royalty did give, especially Henry III, but in Italy it was often the great banking houses who were the major patrons. Across Europe, local arrangements were made both for building and sponsorship. Fund-raising was done in some ways much as we might do it now, including putting collecting boxes in local shops and staging events to encourage visitors and local townsmen to give. One of the more enterprising campaigns was run by the monks of Crowland Abbey, near Peterborough, when the abbot, Joffrid of Orléans, decided to start a major overhaul of building works in the early twelfth century. The monks travelled to Cambridge, where they hired a barn and gave public lectures, which people paid to attend. These proved so popular that they went a long way towards funding the project.[27]

However, the bishop and other worthies were expected to put up a lot of money themselves. Sometimes, as at Beauvais, as much as ten per cent of their personal income for up to ten years, so they were substantial givers. The bishop would then ask the Chapter's canons to do the same. The bishop of Auxerre also personally gave generously to his new church; in the first year of building alone he gave 700 livres of his own money.[28] We have already seen something similar happening at Chartres where they were also able to capitalize on emotions running high to get local people to give. That said, although it was not an insignificant sum, it was not nearly enough. The amount given by the bishop was often so great that he might be seen as the person who built or commissioned something even though the whole project was so costly that, in fact, his donation was a mere drop in the ocean. This level of personal giving was not always going to work and, although bishops were hardly on the breadline, it was a severe burden on some of them. There is an amusing extract from the chronicle of a monk of Durham concerning the building of the cathedral in 1099:

> [Bishop Ranulf Flambard] concerned himself with the work of the church with greater or less energy according as money from the offerings to the altar and burial fees flowed in or was lacking. For with these funds he had built the nave of the church, with its outside walls, up to the roof. But his predecessor, who began the work, had laid down as a principle that the bishop should build the church out of his own income, and the monks should build their monastic offices out of the church offerings. Which rule died with him.[29]

This is revealing because it shows that if the Chapter wanted a new-build to happen, one option might be to find a bishop who had the means and the will to underwrite it. Parishes were also expected to contribute, which still happens today (although the parish share or quota is now needed for more run-of-the-mill items such as clergy pay and pensions and financing the general administration of the diocese). When major works are needed to restore or repair a cathedral now, special appeals and schemes are set up, although the Dean and Chapter would be forgiven for hoping that a wealthy man with the medieval notion of buying a place in Paradise might come along. Our ancestors had no difficulty with this, not least because it was blatantly encouraged by the Church. They were fortunate in that people did believe that there was an after-life and that there would be Paradise (or Hell) and this gave them something to aim at, so to speak. It also helped alleviate some of the misery of their daily existence because it identified a purpose: the worse your suffering, the greater your reward. They believed in the existence of Purgatory, which was a sort of halfway house between earth and eternal bliss. It gave the believer a second chance to atone and the worse their sins had been, the longer they spent in Purgatory and the more unpleasant a time they had there. Because of this belief it was common for indulgences to be granted that would reduce the time believers had to spend in Purgatory. We might wonder how anyone could be so sure of the workings of God's mind that he could write with such confidence that 'he who sows with almsgiving shall reap eternal life'. From what the Cardinal Hugo said in his indulgence letter of 1253 concerning giving to the building of Strasbourg cathedral, it seems that any donation to the work would fast-track donors to Paradise. Other letters of indulgence were more specific and guaranteed perhaps six weeks off the time they would have to spend in Purgatory were they to help repair a bridge or mend a road, both considered pious acts because they helped pilgrims.

Bequests and wills were a good source of revenue. Wills were often written as a knee-jerk reaction to the arrival of the Grim Reaper and, if they were written as opposed to spoken, frequently dictated to a cleric, who was probably the best-educated person available and someone who could write. One hesitates to suggest that any of them took advantage, but it was a good opportunity for a churchman to remind the deceased-designate that he was right on the brink of meeting his maker, so now might be a good time to be friendly towards the church in cash terms. As François Icher put it: they enjoyed a privileged position in the imaginations of those who knew they were soon to die.[30] In fact, wills were often made in the last hours of life, so then it was very real and perhaps not so much the administrative exercise it might be today. In these wills, the person would bequeath his or her own soul to the Almighty and the saints, would sometimes specify what monies should be spent at the funeral, and would often request burial in a particular part of a church. Sir Geoffrey Luttrell of Irnham, Lincolnshire was a case in point. When he died in 1345 aged at least sixty-nine, he left 500 marks for twenty chaplains to say Masses for his soul for five years, forty shillings for clerks to recite the Psalms on his behalf and he gave his war-horse as a mortuary gift to the church. In addition, he ordered an amazing £20 worth of candles to be placed round his body on the day of his death. The majority would have been what are called 'pavement candles' of great size and would have been used in the church for the following weeks, if not months. In effect, Sir Geoffrey was paying the church's lighting bill in advance. Each mourner was to receive a penny and a place at a funeral feast – and this was just what was to happen at his parish church. He also left money to the cathedral in Lincoln and a gigantic £200 to the poor.[31] His legacy to us is his fabulous Luttrell Psalter, now in the British Library, so let us hope that he did indeed get his reward in Heaven. He may have hoped that he would since a century earlier the pope had decreed that there would be a relaxation of a year's penance granted to penitents who gave to the fabric of the then new church being built at Westminster, so his munificence to his local church must surely have been worth something.

Some cathedrals found that they were the victims of overgenerosity in terms of goods. Priests to this day would probably prefer that their church be given money rather than a specific item and the Chapter and Workshop Committee at Milan were no exception. They found that they were continually being given small things for the new cathedral and the solution they came up with was to hold regular auctions to turn the gifts into cash. This was a difficult area because the givers might understandably have felt that their presents were being slighted, and there are more complex issues to do with the motivation for placing a gift within a church, so the clergy had to develop diplomatic skills.

Another way to raise money was to send relics on tour. This might not find so much favour today, although many people would be interested to see the sometimes ornate and beautiful reliquaries. People still flock to major church art or illuminated manuscript exhibitions, but they are usually going for reasons of culture, not spirituality. In the Middle Ages, the intention was different, although the idea of setting up an aesthetically pleasing and inspiring display was the same. It was a very common practice and, by taking relics to the people, monks hoped to attract pledges to the cause, if not actual cash on the spot. If the relic was important enough, donations were likely to flow in as the benefit was twofold:

1. Believers had the chance to get physically close to a saint.
2. Believers were contributing to the place that would eventually house a saint and this, in turn, would encourage that saint to look down on the believers and listen favourably to their pleas.

People really did believe that the closer they got to the remains of an actual saint, the more likely he or she would intercede on their behalf. This meant that their chances of getting into Paradise – or, at least, having their time in Purgatory reduced – were vastly increased. There were plenty of people who truly believed that the main reason they were in this life was to prepare for the next, so this was a genuine and rational response to a genuine and rational way of fund-raising. However, there were several difficulties with this, essentially concerning human nature.

Everyone (it seemed) was building and certainly everyone was trying to raise money, so opportunities were jealously guarded. If the relic of cathedral A was being paraded in the diocese of cathedral B, then B was not going to get that revenue. As a result, some bishops would not allow relics from another diocese to cross into theirs, and there are instances on record of legal action being taken to impose boundaries and to restrict access to some parts of a diocese. That said, some relics toured other countries; relics from Laon went on a fund-raising trip to the south of England and were on the road for about seven months, in total.

Another ongoing problem with relics were conmen, some of whom dressed up as clergy and displayed any old tat in the hope of collecting revenue that, needless to say, never found its way into the coffers of the church in question. They might have had a piece of bark, which, they claimed, was a fragment of the True Cross; they might also have had a fingernail, lock of hair or a tooth, which were all relatively easily obtained, or a piece of bone, which was usually from an animal. That said, there was so much anxiety about what we would call 'body-snatching' and the selling off of body parts as relics that, in 1300, Pope Boniface VIII published a Bull forbidding the mutilation of corpses. The Medical Faculty of Paris formally stopped surgery for the same reason, which was well-intentioned but which put medical knowledge back several years.[32] Not all conmen were freelance, however, since the profession of pardoner thrived: pardoners were licensed to sell papal pardons ('indulgences'). Some of these people were genuine, even if their relics were not. The medieval man in the street was not dim-witted and often saw through these charlatans, as this description of just such a pardoner makes plain in Chaucer's Prologue to the Canterbury Tales:

> A voice he had that bleated like a goat.
> No beard had he, nor ever should he have,
> For smooth his face as he'd just had a shave;
> I think he was a gelding or a mare.
> But in his craft, from Berwick unto Ware,
> Was no such pardoner in any place.
> For in his bag he had a pillowcase
> The which, he said, was Our True Lady's veil:
> He said he had a piece of the very sail
> That good Saint Peter had, what time he went
> Upon the sea, till Jesus changed his bent.
> He had a latten cross set full of stones,
> And in a bottle had he some pig's bones.[33]

Chaucer's pardoner has a cross made of latten; as latten was a material that looked like gold, but was considerably cheaper, this was all part of the con.

Another fund-raising ploy was to persuade someone to donate something significant, such as a whole chapel, so that prayers could be said in perpetuity for his or her soul. Or it might be a set of statues or a stained-glass window. At Chartres, several windows were given by guilds on behalf of their members' souls and at Le Mans, the locksmiths and vine-dressers pledged a window of five lancets in which they themselves are depicted. Patronage will be considered in more detail in Chapter 8.

At every opportunity, the Church hammered home its message: hoard and go to Hell; give and go to glory. A visible outcome of this is that if only one or two of the seven Deadly Sins are portrayed on church buildings, the one most often sees is Avarice. Avarice, being the love of money, is not actually a Deadly Sin (greed is). Avarice is also possibly one of the least interesting sins to carve since it usually only takes the form of a man or woman clutching a money-bag. That said, in almost any line-up of the damned on a Last Judgment tympanum, Avarice is there and good people could save themselves from that sin by giving as much as possible to the Church. Institutions, such as cathedrals, obviously could not be guilty of avarice; just secular types who might be involved in profit-making businesses. As an aside, in that same line-up it is very common to see a prince and a bishop, and it must be said that bishops did not fare well in medieval art unless they commissioned it.

If the money ran out, building stopped, sometimes for days, weeks or even years, which made it difficult to carry on as unfinished stonework and carvings will deteriorate when left exposed to the weather. This was found to be only too true at St Albans in 1195. We are told that in this instance the Master Mason was a deceitful and unreliable man, albeit skilled at his craft. The costs escalated and the walls were left uncovered in the rainy season, causing irretrievable damage to the soft stone being used. The workmen departed in despair, their wages not paid to them.[34] This caused grave problems for the next man appointed, and it is to his credit that he solved them.

The death of a Master Mason could almost bring about the death of the project unless it was properly managed by the Workshop Committee. His plans were their property and they then had to begin the whole process of finding a new Master Mason. This could take months, during which time nothing would be done and the workforce would be obliged to disperse to find employment elsewhere. Times being what they were, the death of a Master Mason was a common problem and an area covered by the Statutes of Ratisbon of 1498 (of which more in Chapter 6). A particularly nicely turned clause says that:

> If a master dies before completing the work he undertook and if another master takes it up, he must complete it and not give it to a third person; this is so that those who commissioned the work in question are not involved in excessive expenses that would be harmful to the memory of the deceased.[35]

All the things discussed in this chapter will have been coordinated by the Master Mason as well as recruitment and manpower problems, which have yet to be considered. These Master Masons were remarkable men, so when, in 1178, one such as William of Sens fell to his death from the scaffolding in the choir of Canterbury cathedral, it was not just the loss of a skilled craftsman, but also of a charismatic leader.

CHAPTER THREE

THE MASTER MASONS I

What kind of men were the Master Masons?

THE INDIVIDUALS WHO coordinated the men and materials must have been extremely impressive characters. It is almost impossible for us in our automated age to imagine how great a physical labour building was and how much encouragement or incentive the workforce required. The volume of noise would have been colossal, making it difficult to brief and check a junior craftsman on site; the number of voices and tools continually in action would have been deafening in the days before ear defenders, so the workforce had to adjust or go deaf. Some of the men would have been working in what we now might see as all but impossible conditions, barely protected from the weather, inadequately shod and clothed, operating hundreds of feet up as the building progressed with little regard for health and safety. Building sites were dangerous places and not for the faint-hearted since every tool, stone, beam, pinnacle and window-frame either had to be winched up manually, carried by one man in a hod, or by two men using a sort of stretcher. Surprisingly, wheelbarrows do not make an appearance in records until the thirteenth century, although that does not mean that they did not exist.[1] One wonders if they were popular among the workers at first because they could be operated by one man, whereas two would have to be employed to carry materials on a board. As a roof boss in a vault could weigh over 200lb (90kg), placing it was not only risky and highly skilled, but also physically tremendously arduous. While the Master Mason delegated to Under-Masons, Master Carpenters, Master Glaziers, and so on, he himself had to have a presence and credibility that would make him instantly recognizable in the cacophony of the building site. Numerous accounts reveal that he wore a distinctive robe to show his office and that this garment might be renewed at least annually as part of his contract, if not more often. There were also rough dress regulations on the building site: contemporary illustrations show that individual workers wore identifying items of clothing, such as knee stockings for apprentices and coloured head-bands for others (a white one for a sculptor, for example). The Master Mason would have known where each individual trade was centred, because a small but crucial part of his role was to plan the building site in the sense of who went where to work. Some trades needed more space in which to work than others and some needed fire or water in quantity. Kilns were necessary just to make solder, for example.

Managing a large workforce

The Master Mason commanded a sizeable workforce, which varied according to the building and the stage it was at. One of the better sets of accounts that survive are those not from a cathedral, but from Kirkby Castle, and they show that 706 mason-weeks were worked in the labouring seasons between May 1481 and November 1484.[2] In the two years beginning 1268, at Beaumaris Castle, there were 1,630 workmen, 400 masons, thirty blacksmiths or carpenters as well as 1,000 unskilled labourers and carters.[3] At Westminster, the number of masons per week were listed between 1292 and 1294 and the numbers ranged from forty-two to ninety-four as well as fifty-five stone-cutters and six carpenters.[4] This does not tell us about the numerous other trades involved and we must bear in mind that while building sites generally closed over winter, this was not always so and sometimes men were being paid, but at a lower rate during the traditional down-time from All Saints (1 November) until the Feast of the Purification (2 February). Not all of the men involved were skilled enough to have been held on a retaining fee. It was not until relatively late in the period that stone cutting was done on any kind of production line, which happened for the first time at Amiens cathedral in the thirteenth century and which speeded things up. In essence, rather than have every stone cut individually to fit the space needed, they turned the equation around and decided to make the cathedral out of identical stones wherever possible. This meant that they could carry on cutting throughout the year. It was not possible to carry on laying the stones, but it did mean that a ready-cut supply could be stock-piled during the winter. It is surprising now to realize that this was not done before, although there is evidence that something similar may have occurred at Durham in the 1090s.[5]

It is also all but impossible to track casual labourers, of whom there were thousands. There are cases when a mass workforce (including women) was hired to get a job done quickly, such as in 1302 at Linlithgow, in Scotland, when

103 ditchers were employed at tuppence each, with no fewer than 140 women
working with them at a penny a day. In one week in 1253, wages at Westminster
Abbey were paid for 'one hundred and seventy-six inferior workmen' with overse-
ers, clerks and two horse-carts, which came to a total of £9 17s 2d.[6] In the same
week, the following were paid a total of £15 10s 1d (or an average of 1s 10d) to
each workman: thirty-nine cutters of white stone, fifteen marblers, twenty-six
stone-layers, thirty-two carpenters, two painters at St Alban's with an assistant,
thirteen polishers of marble, nineteen smiths, fourteen glaziers and four plumbers
(1s 10d would have been a very high wage at this time, but many of these men
were skilled, not general labourers: marblers, for example, generally commanded
a good rate of pay).

It is not easy to produce a definitive list of all the trades employed in the building of a cathedral or church, not least because such a list would depend on the
dimensions of the building and the level of decoration, but the list in Appendix I
shows over eighty trades and gives an idea of the range of knowledge needed by
the Master Mason as to who did what, if not in detail, then at least with some
general awareness of their requirements and the pace at which they worked. He
also needed to be aware in advance of how great the workforce might be since
skilled men generated numerous assistants and others to support them, and all
would have to be paid, sometimes out of the Master Mason's own wage. This was
a diverse and fluid workforce, all of which had to be found, recruited, trained,
coordinated and administered by one man, so although he had the same ability
as the modern architect to design a building, it may be that today there is not
the same need for sheer force of personality. The Master Mason was, of course,
working without the aid of calculators and computers, but was undoubtedly also
less hampered by bureaucracy. The main difference between the medieval Master
Mason and the modern architect is that today he is not necessarily the project
manager whereas his medieval forebear most definitely was. It is these men who
are the subject of the next two chapters: who were they and what were they like?

Who helped the Master Mason build a cathedral?

To find these extraordinary men is less difficult than one might think. There is
a common belief that the Master Masons and general workforce of the Middle
Ages were anonymous, usually by choice, so that their work might truly be solely
for the glory of God. While this may have been the case for some, the names and
some insight into the lives of many have come down to us; in fact, our medieval
architects are better known to us than musicians of the same period, and are at
least as well known as our painters and sculptors.[7] For this information, I am

indebted to the late Dr John Harvey and his remarkable biographical dictionary of *English Mediaeval Architects*, on which I have drawn a great deal. Unless otherwise specified, the information about individual Master Masons will have come from his book. Harvey names 1,300 individuals who operated between AD 968 (Ednoth of Worcester) through to 1550, and has taken those men who can be said to have designed all or part of a building and/or who supervised construction and the building site, no matter what his own professional background might have been. It is natural to assume that the Master Mason was actually a mason by trade and training. Of Harvey's list, 1,284 were active before the Reformation and it is true that the majority began their professional lives as masons, but only just, being 693 or fifty-four per cent of them. Of these, it is not clear how many were originally stone-cutters, rather than masons. The next largest group are the carpenters at thirty-two per cent, which is not surprising since the senior carpenters and masons had to work together very closely and have a good understanding of each other's work. To bring it to its most basic level, the main two ingredients of any building of that period were stone and wood.

In Appendix II there is a breakdown of trades from which Master Masons came. This excludes those men who were also monks, priests or clerics. Only one of them was referred to as being an architect in his own lifetime, a term we now use for Master Mason (although the modern architect's training and background would be rather different, as has been indicated). This pre-Reformation architect was Isambert de Xaintes (Saintes, in the south of France), who was a scholar with a reputation for designing bridges and was summoned to England in 1202 by King John to complete the work begun on London Bridge by the only Master Mason priest, Peter of Colechurch, who died three years later. As this priest is first heard of as an active bridge-builder in 1163, he seems to have had a long career and may have been of a considerable age when he died. This may also account for why someone else was brought in to finish the job.

There were a number of carvers, engineers, joiners and marblers among the ranks of the Master Masons because these were the skilled trades that required a man of intelligence and aptitude, not just with his hands, but also in appreciating the best use of time, space and materials. The same was true of goldsmiths, tombmakers and brass-workers, but they were a smaller pool from which to draw in the first place. None of the Master Masons were glaziers by trade, which was a job demanding great skill, but they could be said to have been more concerned with art than artefacts. Remarkably, four Master Masons started their professional lives as bricklayers and brick-makers, which were trades deemed to require less intellect, so they must have been men of exceptional drive and charisma.

One of these two bricklayers was a man named Richard Plantagenet, who lived between 1469 and 1550, when he died at Eastwell in Kent, possibly aged over eighty. It has been claimed that this man was an illegitimate son of Richard III and that the king acknowledged him the day before he lost his life on the field of Bosworth in 1485. Following his father's defeat, he is supposed to have realized that discretion was the better part of valour and thereafter made his living as a bricklayer – a strange choice for someone who wanted to blend into the background since we are also told that he was well educated and fond of reading Latin. That said, it was probably a job that required no credentials and, therefore, no questions to be answered, but, if he was really given to sitting apart from his colleagues so that he could read poetry in Latin and French, it is remarkable that no one noticed this until 1545 (sixty years after the battle and when Richard would have been a very old man himself) when Sir Thomas Moyle is reputed to have persuaded him to tell his story. Another point is that the poetry reader was evidently still a common workman at the time of his discovery, not a Master Mason. While there is no reason at all why a tomb at the ruined church of Eastwell, near Ashford, should not belong to one of Richard III's illegitimate children, he might not be one and the same person as the bricklayer-cum-Master Mason, but happily that will never be proven or disproven.

It is not hard to see how John of Limoges grew to be a Master Mason in 1282 because he designed tombs, some of which were small buildings in themselves. It is less easy to see how John Orchard came to be categorized as such since he was a brass-worker in the last quarter of the fourteenth century making (for example) angels for Queen Philippa's tomb in Westminster Abbey. Whatever his achievements, they were certainly lucrative because he owned several properties in London by the time he died. Some men were specialists, such as Richard and John Bochor of Pluckley, in Kent, who built mills, notably one at Chartham for the prior of Christ Church Canterbury in 1438, for which they were paid twenty-two marks (£14 13s 4d). Twenty-three Master Masons (1.8 per cent) were originally skilled in more than one area and that additional expertise must have made them even more of an asset, whether their talents were similar (such as the carpenter/joiners) or different (as in the case of William, the engineer and mason, working at Durham in 1195–9).

Plenty of other Master Masons were also multi-taskers, not least the religious group (one likes to think that they remained attentive to their calling), but several had other professional interests and it may be that these were really their main careers, with their work on building sites coming second. From his will, it seems that Richard Lynne who died in 1341 was mainly a sheep farmer with land at Enfield and Brixton. Some second careers were building-related, such as Thomas Danyell (active 1461–87) who was a mason and a stone merchant, and William Wright, who was both a carpenter and a timber merchant in the latter part of the fourteenth century. He worked at Ripon Minster, where he supplied timber for the towers and carried out numerous works ranging from hanging bells, making doors and a rood-loft for a Ripon church, but is also recorded as having made the high vault of the nave of Thornton Abbey, in Lincolnshire, in 1391–2. Sometimes it is less clear in precisely what additional ways the Master Mason was gifted, but it would seem that Lando di Pietro of Tuscany was what would now be known as a 'good all-rounder' because it was said of him at the time that:

> Master Landus, the goldsmith, is an experienced master not only in his own above-mentioned craft but also in many other arts, he is a great man, subtle and ingenious in devising new things, be it for the building of churches or for the building of palaces, or houses of the commune and roads and bridges, fountains and all other public works ...[8]

Getting on in the building trade

What we do not know is how the skilled craftsmen graduated to the rank of Master Mason. No record survives that shows that important step of the first commission or any type of ceremony. In reality, there was probably no set procedure; it seems likely that luck as well as skill, experience and being in the right place at the right time was involved, as it still might be today. A would-be Master Mason would have needed to have demonstrated his ability within his trade and presumably he gradually stepped ahead of his peers as he matured in age and proficiency. Common sense suggests that he was probably offered the work of Under-Master Mason to begin with, or perhaps a minor project, but we do not know. Although this book is about the building of our medieval cathedrals and churches, it should also be said that Master Masons and the workforce took employment where they could, so some of the evidence that survives shows them working on town walls, individual houses, university colleges and the like, but the principles of building and organization were similar whether the Master Mason was Henry Yevele planning Edward III's tomb in 1377 or, less glamorously, Geoffrey of Carlton working on the canons' latrine at Windsor Castle in 1352.

While dispelling myths, it is occasionally assumed that fathers followed sons into the same business in an endless succession until a less family-centred age was reached, possibly quite recently. In fact, then as now, this was not the case, although we cannot know how much store a father set on his son taking over his business. No doubt they hoped that this would happen just as any modern father running his own business might hope for the same. There were numerous Master Masons with the same surname who may well have been related such as

Hugh and Robert, both of Albemunt (also listed as Abbemund, Abbendon and Ablemond), both carpenters in the early thirteenth century. A blood relationship seems likely, though there is no evidence for it; equally, there is no reason why two men from the same place should not have trained as carpenters and become Master Masons.

Out of the 1,284 names under consideration, only eight fathers and sons acting as Master Masons and coming from the same profession have been confidently identified. In chronological order, these are: Ailnoth and Roger Enganet, engineers 1157–1216; William and William Ickenham, carpenters 1388–1424; John and John Ickenham, carpenters 1395–1414; John and John Massingham, carvers 1409–78; John and John Bell, masons 1418–88; Thomas and Thomas Redman, masons 1490–1536; Humphrey and Roger Coke, carpenters 1496–1545; and Richard and Roger Rydge, carvers 1499–unknown date. It seems likely that William and William Ickenham and John and John Ickenham were related, and Robert, Henry and Robert Janyns, masons operating between 1438 and 1506, are thought to have been father, son and grandson.

Inevitably, other sons did better than their fathers, such as the carpenter Richard Fitzwiching, who became a Master Mason while his father did not, so there will have been numerous families associated with the building trades who are invisible to us because not enough of the family members achieved particular prominence. At the other end of the scale, there were whole families of Master Masons, notably the Ramseys, Yeveles, Vertues, Hermers, Herlands and Worlichs, while the remarkable John Wastell, whom Harvey described as 'perhaps the most significant figure in the last age of Gothic architecture in England', was probably the son of a weaver. This is a natural progression and it is actually remarkable that single families could produce more than one man capable of being a Master Mason (one wonders what life was like for less talented siblings), and there is the rather touching story of Ralph Burnell, carpenter and Master Mason for Henry III between 1220 and 1262. He was well paid and held in high regard, being also a King's Carpenter who had worked at Windsor Castle. When he died in 1262, his son Thomas was appointed in his place, which suggests that Ralph continued to hold office and work until his death, so he must have reached a good age. Though Thomas had not been trained as a carpenter and so could not carry out the work, Henry III, 'in regard for the long service of Ralph granted Thomas the fee of three pence a day for life of his special grace'. The appointment and the pension says a great deal for the ability of Ralph, the affection in which he was held and the general esteem the Master Masons commanded, which will be looked at more fully later. This was a significantly good pension for young Thomas, although we do not know if he used it well.

Masons' marks

Before considering how we might know the names and some of the lives of these men, it is worth mentioning the way that people erroneously think that they know them and that is through masons' marks, also known as bankers' marks, thousands of which survive. Marks that appear to be similar to each other can be found in cathedrals and churches across Britain and Europe, but just because the marks are sometimes alike, it does not mean that they were made by the same man and nor can we usually attach a name to a particular mark. Because the marks are simple in design, it is easy to think that we see connections between some of them, but there is no evidence that a worker used the same mark at different sites. At Beverley Minster alone 700 marks have been found on the nave piers, showing that more than a hundred different men worked on them, although it is not known if this was continuous or if there was a break in the work.[9] The bankers' mark is not so-named for the treasurer's benefit, but because these were skilled men working at a bench (from the French, *banc*) cutting stone into shape, as opposed to being the lesser-skilled men who set the shaped stones into place. Some marks found are, in fact, for positioning, not for identification, such as the ones now exposed on a half-timbered building in Tombland Alley in Norwich (and these can cause confusion if the item being positioned has subsequently been moved, as has happened on the façade of Rheims cathedral). There is a possibility that some may be marks of provenance to show which quarry provided the stone, but that would be rare since a Master Mason would use stone from the same place as much as possible to keep the overall look uniform. Most ordinary masons and other workers were paid by piece work, so they would carve their mark onto the work that they had done as proof to show whoever was in charge of the money, arguably another reason why the Master Masons cannot be found in this way because they were more usually paid a salary or a set fee.

Many masons' marks, carved into stone, have survived. It is rare, but not impossible, to find them elsewhere. The most obvious examples of ones on wood are those carved onto the misericords in the stalls of Exeter cathedral and the church of St Lawrence at Ludlow. The misericords at Exeter are the oldest complete set in Britain, thought to date back to 1238–44. Some are exquisitely carved and one, featuring a crowned knight fighting a leopard, also has a distinctive pattern of two roundels with a dot in the centre, which is thought to be the mark of one of the carvers. At Ludlow, seven misericords dating to 1389 display a stylized three-branched twig on the left side, which is also thought to be the maker's signature. Cathedrals and churches are often littered with these marks, some of them clearly showing a particular worker doing several courses of stones before someone else took over, the reasons for which have been lost in time. When there are only a few marks, it suggests that the masons were not paid by output but received regular wages, as at Salisbury cathedral.

Despite all this activity, the reason we cannot say exactly who carved the marks is simply because no record has survived, if one was ever kept. The identification of work was not needed in the long term; the ownership of the marks was noted for payment on site at the time; they had no further function and were frequently painted over. They are usually simple so that they could be etched quickly and might have been based on an initial. If so, a mark could equally have been adopted by a junior mason taking the initial of his immediate boss. It may be that the leading mason was paid for all the work of his team, which he then shared, so the same mark might refer to many workers. No doubt individual masons had a particular mark they preferred but, if it was too similar to someone else's when he went to a different workplace, then he would have had to change it and, likewise, very little survives showing how they were allocated or any rules surrounding them. Indeed, one of the few references comes from the Statutes of St Michael, drawn up in 1563 in Ratisbon (modern Regensburg), of which article 55 states:

> And no one must by his own will and authority change the mason's mark that the corporation has given to him. If he wishes to change it, he may only do so with the good will and approval of the corporation, which must have been informed of it.[10]

Plainly, he needed to inform the authorities of his mark or he might not get paid. On the Continent some rules were laid down in the Torgau Statutes of 1462, but these were also drawn up long after most marks were made and there is no reference to them west of Germany. A singular exception to this is the case of Mathes Roriczer of Regensburg. Hans Holbein the Elder drew his portrait in 1490, named it and included what must be a mason's mark. Mathes did begin life as a mason and a very similar mark has been authenticated to his mason father, Conrad. It seems likely that Mathes took his father's mark and adapted it slightly to make it his own.

Jennifer Alexander, who has written about the marks at Beverley Minster, also points out that had they been intended as a permanent name-tag, so to speak, they might have featured in other records such as accounts or contracts or have been shown on memorial plaques because that would have been an obvious place to link the man to his work had it been required at the time.[11] Of his 1,300 Master Masons, Harvey is only able to assign marks to three men, all very late in the period: Thomas Drawswerd 1495–1529, John Orgar 1508–46 and Richard Lynsted 1522–42. These marks, which take the form of etched bankers' marks, have been found drawn on documents. Drawswerd's might not even qualify because it is described as a merchant's mark and so must have appeared on documents and may not have been used on buildings at all. It is certain that this man could write because he held numerous civic appointments. Lynsted's mark was beside his signature, so we assume that he could write; Orgar's was in lieu of a signature and is dated 1536, so he may not have been literate. It seems likely that most Master Masons could read and write, given the intelligence level needed to achieve that status. Intriguingly, at Beverley Minster there is a mark in the style of a rough cross that has the word 'Malton' carved beside it. As there was indeed a William de Malton who was Master Mason at the Minster in 1335, it may seem that, at last, we have definite evidence. Sadly, this is almost immediately confuted by the fact that 'Malton' is incised twice and by two distinct hands, although that may not be relevant in itself as William de Malton could have instructed someone else to do it for reasons known only to him. That would raise questions as to why he felt the need to have his name spelled out so specifically in such a small area and not, apparently, elsewhere. In addition, we have no idea if the words were written at the same time as each other, never mind as at the same time that the masons' mark was made; they could be the graffiti of any passing Maltonian over the centuries. Happily, there are several ways to find the Master Masons: inscriptions, wills, contracts, accounts, tax assessments and lawsuits, to name but a few.

Portraying Master Masons in stone

Although no masons' marks have been found recorded on inscriptions on tombstones, it is not at all unusual to find Master Masons identified by images of them with the tools of their trade; there is an intriguing reference to Walter Dixi of Barnwell, Cambridgeshire, who made a grant of land to his son, Lawrence in 1277, 'sealed with his device of a hammer between a half-moon and a star or sun'. Master Masons' tools seem to have been their own personal property and very much part of their badge of office. Where an inventory has survived we can see that junior masons and labourers were issued with tools on a building site, but the particular items such as set squares, levels and plumb lines used by the Master Masons were never listed and so must have been part of their own private apparatus.[12]

Tombstones with images of Master Masons are rare, but one may be seen at Crowland Abbey, where the image of William de Wermington, who died c.1350, is shown with his compasses and set square (*see* plate 6). In Lincoln cathedral's cloisters can be found the tomb of Richard of Gainsborough, who died c.1350; he too is shown with his set square. The tombstone of Hughes Libergieres has been fixed to a wall in the north transept of Rheims cathedral so that his finery can be examined (*see* plate 7). Hughes holds a model of a church, has a set square by his right foot and a lewis (lifting device) by his left. There are a few portraits, perhaps most famously the bronze head of Lorenzo Ghiberti on the east door of the baptistery in Florence and Peter Parler at Prague. In England we see portraits of William Wynford, Hugh Herland, Simon de Membury (clerk of works) and Thomas the glass-painter in the east window of Winchester College Chapel (albeit an 1822 copy of the fourteenth-century original), but again, these are rare. One of the heads on the north-west arch of the Octagon at Ely cathedral seems likely to be one of the many Master Masons working there over the years; the date of 1322–8 would, in turn, make this likely to be the one known as John the Mason. William de Ramsey III, one of the prolific Ramsey family, is probably the subject of a capital in Lichfield cathedral. In Westminster Abbey, a finely carved head in the north transept is thought to be either Henry de Reyns, who originally designed the building and died *c*.1253, or the man who took over from him, John of Gloucester (this is the successor to the church built by Edward the Confessor).

It seems very likely that some of the heads carved on corbels are portraits of Master Masons. There is one in Wells cathedral that may well be Adam Lock, who died around the year 1229 (*see* plate 8). He is often credited with designing Wells' lovely west façade. Unfortunately, he is likely to have died before that happened, although he will have been involved in creating the western end of the church. A craftsman's head in Lincoln cathedral may show Richard the Mason, who was in charge of the works around 1195. Exeter cathedral boasts a carved head in the crossing, which is probably Roger the Mason (died 1310). Henry Wy, the Master Mason of St Alban's Abbey in the early fourteenth century is likely to be the man commemorated along with an abbot, a queen and a king, while Henry Yevele is carved on a boss in the east walk of the cloisters of Canterbury cathedral (*see* plate 9) and this representation was probably copied from a death mask (he died in August 1400, possibly aged eighty). It is frustrating looking at these images since one is so close to the carvings, but so extremely far from discovering any solid facts about whom they may represent.

Three Master Masons are shown in the centre stone of the labyrinth in Amiens cathedral in Picardy along with the bishop, Evrard de Fouilloy, who had laid the foundation stone in 1220. An inscription round the edge tells us that the Master Masons were Robert de Luzarches with Thomas de Cormont, who directed the work and then his son, Renaud.[13] (The labyrinth was reworked in the nineteenth century.) This was a real honour: the centre of the labyrinth represented Jerusalem, so these men are shown as having reached the holiest of lands, presumably in the after-life. So portraits do exist, but at this distance it is not possible to identify them with certainty unless they were clearly annotated at the time. Sadly, this rarely happened, either because it was not the fashion, or it simply was not needed because they were so well known at the time. One of the problems we have, looking back into history, is that it is often plain that our ancestors could have no idea that their works could survive for as long as they did, and perhaps they did not imagine that we would be interested. History in those days was seen either as a story of kings or as a way of conveying a moral point (good will overcome, bad will be defeated), not a record and analysis of events. People were not interested in knowing how their ancestors conducted themselves, being more concerned with improving their own lot.

Other clues to finding Master Masons

There are other inscriptions and clues on display in our churches, should we wish to look. The parish church of Bridekirk, in Cumbria, has a delightful font with scenes of the baptism of Jesus and possibly an expulsion from Eden, some magnificent monsters and, beneath, an inscription, a small mason energetically brandishing the tools of his trade (*see* plate 10). Harvey dates this to around 1150, but there is much debate about it. The crude Romanesque style of decoration with its mixture of the grotesque and human is seen on twelfth-century churches as far apart as St Margaret's in York, Wissett in Suffolk and Barfrestone in Kent. However, the inscription running above the Bridekirk mason's head is in runes, causing some to believe that it is either much older or that runes were still being used much later than is usually thought. Runic inscriptions are less straightforward to translate and this one has at least two main variations. In 1816, two antiquarian brothers, Daniel and Samuel Lysons wrote that the inscription read:

> first satisfactorily explained by Bishop Nicholson, in a letter to Sir William Dugdale. He read it thus: *Er Ekard han men egrocten, and to dis men red wer Taner men brogton*; i.e. Here Ekard was converted, and to this man's example were the Danes brought.[14]

This interpretation suggests that Ekard was a prominent local figure who became a Christian, leading others to do so and probably endowing the church. The word 'Danes' at this time denotes any Scandinavian or Viking, not necessarily someone from Denmark. However, both Harvey and M. D. Anderson believe that the runes say '*Rikarth he me iwrokte and to this merthe gernr me brokte*', which has been interpreted as 'Richard made me and to this beauty carefully brought me'.[15] There is a replica of the font in the Victoria and Albert Museum, in London, allowing researchers easier access to it.

Some inscriptions are more straightforward: *Robertus me fecit* ('Robert made me') is inscribed on a carved capital in Romsey Abbey, in Hampshire, probably dating to 1120–50; *Godfridus me fecit* at Chauvigny; *Durandus me fecit* on a roof boss in Rouen cathedral – Durand being the man most likely to have been the Master Mason of the nave from 1214–51; and *Alexander me fecit* is clearly incised on another font (this time at Tiköb, in Denmark). Alexander was an Englishman working overseas in the latter part of the twelfth century. Another font with an early inscription is at the church of All Saints, Little Billing, in Northamptonshire. It is a slightly clumsy-looking thing, but the writing is easy to see even after all this time, although a part of it is missing because of a repair. What we can see reads:

> *Wigberhtus Artifex atque cementarius hunc fabricavit*
> *quisquis suum venit merger corpus procul dubio capit*
> ['Wigbertus, craftsman and mason made this
> Whoever comes to immerse his body will doubtless gain']

By 'gain', we may assume that the writer meant whoever had been baptized would have gained spiritually through their baptism. This font is thought to date to just before the Conquest, possibly to 1050.

There are a number of inscriptions, not only in cathedrals, but also in parish churches, where one might expect a Master Mason to have wanted to display his name, if only for marketing purposes. *Giselbertus hoc fecit* ('Gislebertus made this') was carved on the *tympanum* at Autun cathedral in France in or about 1130, and at Chartres we see *Rogerus* carved on the twelfth-century frieze of the Royal Portal with no additional information. We assume that this was the sculptor. The positioning of this name is interesting because it is immediately below the scene showing the Last Supper and could be said to be directly below the seated figure of Christ or that of the kneeling Judas. More interestingly, originally it must have been at least partly obscured because a figure standing directly in front of it has lost its head, which adds a further dimension to the character of the carver. Perhaps he was not the main man after all, but someone lower down the scale (albeit literate) who wanted his name hidden among the myriad of figures, or was someone connected with the cathedral in a different way. Again, we are unlikely to find out the truth.

Richard Uppehalle was recorded in 1297 in an inscription as the founder of the work of the east walk of the cloister at Norwich cathedral, or at least of laying a stone when the work began. A stone on the north side of the Chapter House door had the words, *Ricardus Uppehale hujus operis inceptor me posuit*, but only the name is now visible. William the Geometer, a mason, asked for prayers to be said for his soul on his tomb slab, which was found in the chapter house of Bristol cathedral and is now in the north choir. It is not clear if he was the Master Mason in charge of the alterations to the east end of the elder lady chapel, which took place around 1280, or if he designed the new choir, which dates to about 1298. Robert Day was a carpenter in the early sixteenth century in Cornwall, where he carved the delightful bench ends at St Nona's church at Altarnun. He made a series of the instruments of the Passion, of people working, playing musical instruments and jesting, as well as a wonderful field full of sheep, and he signed one bench-end with 'Robart Daye maker of this worke'. There is reference to a William on the same inscription, which is now too worn to be read.

Lastly, there is Andrew Swynnow, a mason, of whom an inscription says that he began the south jamb of the tower arch of Coton church, in Cambridgeshire, on St Wulfstan's Day 1481. It is likely that he also worked at Caldecote because of stylistic similarities. This is gloriously precise after all the vagueness of other inscriptions, although it probably should be assumed that it refers to the day commemorating the translation of St Wulfstan's shrine on 7 June, rather than his actual feast day of 19 January, as it would have been unusual to start a build-ing project in the winter. It is doubtful that anything can be read into choosing that particular day to begin work: St Wulfstan was the bishop of Worcester and the only Anglo-Saxon bishop allowed to continue in post after the Norman Conquest, which could be taken as a sign of his great success or of collaboration. He lived to the great age of eighty-seven. As an aside, it was St Wulfstan who made the strangely apposite observation:

> The men of old, if they had not stately buildings were themselves a sacrifice to God, whereas we pile up stones and neglect souls.[16]

Men who made fonts seem to be statistically the most likely to have signed them, although it is doubtful if that is particularly significant; as ever, when we look back this far into history we can only analyse the things that have survived. The signing of fonts seems to have been acceptable relatively early on; the three examples in this chapter range in date from *c*.1050 to 1200. It may only be a coin-cidence that it was around the year 1200 that a change in font fashion occurred and they started to be octagonal, rather than round or square.

There will be other portraits and inscriptions, but it must be said that there are not many overall. One wonders how many of the numerous heads found in churches with no identifying symbols or artefacts are actually those of Master Masons and other workers, recognized at the time and probably acknowledged as being in a fitting place. This was still largely an oral society in which stories and facts were passed from father to son, but it took only one father not to tell his son for the knowledge chain to be broken.

Would that they had all followed Humphrey Coke's (or Cook's) excellent example in 1531. He was buried at the Savoy Chapel, in London, which burned down in 1864, but the inscription on his gravestone is recorded as having said:

> Of your Charity pray for the Soul of Humphrey Coke, Citizen and Carpenter of London, and Master Carpenter of all the works to our Sovereign Lord King Henry the Eighth, and Master Carpenter at the building of the Hospital called the Savoy the which Humphrey decreased the 13th day of March in the year of our Lord God, 1530, and lieth under this stone.[17]

In fact, he died in 1531 in our terms because we count the new year as beginning on the first day of January whereas previously years were counted from Spring to Spring. There is also an error on this gravestone as Humphrey Coke was not only alive on 13 March, but made a codicil to his will on the 18 March. It seems likely that he did this on his deathbed and that the thirteen was carved instead of an eighteen by mistake. Another Master Mason appears to have made his own tombstone: John Aylmer, who died aged about seventy-seven in 1548. He had, in fact, been an executor of Humphrey Coke's will, but in his own, which he wrote on 10 July, he specified that he would like to be buried in St Saviour's, Southwark, 'within the Chapel of the Trinitee before the Altar beneath the steppes there as I have laid a marbull stone all redy for my burial with certain pictures.'[18] It sounds as though he did not expect this request to be denied, which is one of many indications of the status of the Master Mason, since only the most esteemed or wealthy parishioners would be buried so close to an altar.

The evidence for Master Masons in wills

Wills are an invaluable source of information, as are contracts and accounts. However, it is in the wills that we see the Master Masons acting as friends to each other that makes them easier to see as real people. Their wives were friends, too. John Clifford died in 1417 and his work on London Bridge was taken over by Richard Beke. The following year Clifford's widow Lettice also died and left a piece of silverware to Alice Beke. John Clifford had been an executor for Henry Yevele in 1400 (along with Stephen Lote, another Master Mason and others), and we know that they had worked together several times over the years. Hugh de Hedon was executor for John Hyndeley in 1407, as was Thomas Mapilton for Stephen Lote in the winter of 1417, so it is clear that these men were more to each other than mere colleagues on a building site, but were firm enough friends to be trusted to look after each other's families' interests when they died.

Lote's will is full of detail. He added a codicil when he was ill and presumably realized that he was dying, and it is this that provides the most interest. After saying that he wanted to be buried in St Paul's churchyard beside his wife, Alice, and leaving money to three churches and a piece of silver to be sent to Hythe church to be made into a chalice, he bequeathed 13s 4d a year for four years to Thomas Mapilton:

> from my shops situate in the great cemetery of the church of St Paul, London,
> in which John Parker and Walter Lucy dwell so that he be friendly and well
> disposed in the making of the tomb of the Duke of York.[19]

Lote was working on this tomb at the time, the Duke having been killed at Agincourt two years before and it is noteworthy that he used his will to hand over the work, so to speak, to someone he evidently knew well and could trust to finish it to a very high standard. Something similar happened in 1424 when the Master Mason, Roger Flore of London and Oakham, contracted with Thomas Nunton to vault the tower of Oakham church in Leicestershire and made provision in his will for this to go ahead should he die in the meantime. We know that Nunton got the five marks allocated to him, but it is not clear if he was the man who carried out the work, or even if it was ever done.

Stephen Lote's apprentice, John Capell, was generously taken care of with a caveat that this was subject to his good behaviour and that he should spend half his legacy for the good of Lote's soul – one wonders if the lad was a bit of a tearaway since two other apprentices who may have been brothers were left one pound each without any conditions. Thomas Mapilton was also left a piece of silver and a bed of 'tapisserwerk' with a coverlet cut for it and 'all and singular necessary belongings that are in his kitchen and room in London … my whole bed in my chamber at Schen [Schene Palace] and all the patterns that be there',[20] which gives us further evidence that designs were physically recorded. Thomas Mapilton was to become one of England's greatest architects, even being called upon to advise on how to vault the cathedral in Florence. Another Master Mason, Walter Walton, was bequeathed 13s 4d a year for four years from the rents of Lote's shops and a piece of silver. These are substantial bequests to his friends, but his wife, sons and daughters were already dead as were his parents; he also left money to those churches whose graveyards accommodated his relatives.

Robert Janyns (junior) and John Lebons were both witnesses to Robert Vertue's will in 1506. In 1513, John Brond was left forty shillings by John de Vere, Earl of Oxford, which suggests that Brond was probably working for the earl at the time since he was generous to a number of his staff. Four years later, Brond's own will showed him to own several houses, and he asked 'My right

Reverende good Lord John Abbot of the Monastery of Bury' to act as the supervisor of his will and also 'to be especially good lord to Thomas my son'.[21] The age of Thomas is not revealed, but this plea seems most likely to mean that when his father died, he would be left a young orphan, although it might also be a way of reminding the abbot of a previous promise to put business in Thomas' way.

In 1256, John the Mason was left three marks (two pounds) by the bishop of Norwich, Walter de Suffield, and we can see the interaction between the Master Masons and the hierarchy of the buildings they worked on several times. Elias of Dereham, who was also a cleric and canon of Salisbury and Wells, was one of the executors for Stephen Langton, archbishop of Canterbury. The sub-dean of Wells cathedral was executor for Adam Lock in 1229, a man whom Elias also knew well. More frequently, the widows acted as executrix, as shown by Sarra, widow of Richard of Gainsborough in 1350. Other such women acting as executrix were: Joan Aylmer, widow of Thomas (1407); Sabine Butler (1413); Clemency Lardyner (1414); Margaret Goneld (1445); Margaret Cony (1446); Alice Goldyng (1451); Margaret Barton (1455); Amy Clerk (1459); Marjorie Couper (1459); Christian Vile (1472); Joan Whately (1478); Alice Redman (1516); Custance Russell (1517); Isabel Swayn (1525); Joanne Redman (1528); and Margaret Drawswerd (1529). Some of these women will have had a great deal to do to sort out their husband's estates, especially Joanne Redman, who was left with a lot of property, but also with detailed instructions as to how to deal with it.

It was common for wills to begin by disposing of the soul, bequeathing it to the Almighty, before the individuals concerned themselves with the souls of others and especially of their families. The Master Mason Henry Redman, the husband of Joanne, was no different in that he specified that he would like to be buried on the north side of the choir of St Laurence at Brentford, where the vestry would be made, so the church must have been undergoing refurbishment. This church was to receive several gifts as well as ten marks a year for a priest 'to synge and rede for my soule and the good helthe of my wife, and the souls of my father and mother and all that I am bound to praie for and all christian soules'. He left money to the convent of Syon and some premises he owned at Brentford were allocated to the parishioners, the income from which should raise £3 6s 8d every year to support the parish priest, which was a generous gesture. These same premises were to remain for the use and benefit of his wife during her lifetime. They appear to have consisted of two cottages and four acres of land, which came under the care of some trustees Redman appointed. He must have had more property elsewhere because he ends his will by saying:

Also to Willm Reedman ... the howse that I dwell in and the lands not bequest ne given after the decease of my wife Jone, if the said Willm Reedman

dye without heyres before my wife Joanne all the lands and howses to give and
to sell at her discretion yff the said Will, Reedman lyve after my wife, then I
will that my dwelling howse with the lands not bequest remayne to the next
male of my blood beyring the name of Reedmans in Huntingdonshire, beside
our Lady of Reedbone [a chapel at Hepmangrove], also I give and bequethe
to my wife Joanne which Joanne I ordeyne and make my sole executrix ...[22]

The Master Mason Nicholas Waleys who died in 1403 nominated his servant
Margery to be one of his executors, the other two possibly being friends since
they did not share his surname. His wife had already died (he requested burial
in the abbey church of St Augustine in Bristol beside her). Appointing a servant
was unusual particularly as he had a daughter, Isabel. We are not told her age, but
assume that she must have been an adult because he made no provision for her
upbringing, but did leave her two of his six shops. One wonders if the servant was
a woman of formidable character! He also gave to his vicar and to the upkeep of St
Michael's in Bristol, these things being likely to help him get through Purgatory.

In their wills, the Master Masons showed a concern for the less well-off and
many of them gave to charity, although this was not quite in the same spirit
that we might do that today; we hope that we will be remembered by those who
survive us, they were hoping their Maker would not forget them in the future.
They were specifically trying to impress the Almighty when they arrived at the
pearly gates with the hope of being allowed through them. Most of them gave
to a church, often to several and sometimes to a specific part of a church, as did
Thomas Ide from Bury St Edmunds, who left money in 1480 to the fabric of
the bell-tower of the abbey. This sounds as though it was work in progress with
which he was either associated professionally or out of affection for a local project,
and is rather reminiscent of fund-raising events today to refurbish an organ or
repair a roof. Numerous other Master Masons gave generously whether it was
to St Paul's cathedral (William Herland) or St Augustine's Abbey, Canterbury
(Thomas Teynham left three quarters of corn to the new work there in 1488).

Even more were determined to be buried in exactly the right spot for the
Almighty's eye to fall on them favourably. John Mapilton, the brother of Thomas,
chose the church of the Carmelites in Fleet Street and asked to be put before
the image of St Christopher painted on the wall, next to the grave of William
Abyndon. He must have known that there was a space there and, since he would
be making his last journey, St Christopher would be appropriate, being the patron
saint of travellers. This also makes it likely that he was buried in the north aisle;
traditionally, St Christopher was painted on the north wall opposite the south
door and so would be the first thing that visitors would see as they entered the
church, reminding them to give thanks for their safe arrival.

In addition, others asked for Masses to be said for them either thirty days after their death and/or on their year's 'minds' (anniversaries). In 1466, John Porter not only left money to York Minster, the church of St Michael le Belfry, Mount Grace Priory, Lincoln cathedral, St Peter Eastgate of Lincoln, and to the guilds of St Mary of Louth and Corpus Christi of York, but also asked for Masses to be said for him to help him on his way. Robert Stowell, who is the only one described as 'gentleman' and 'esquire', also gave money so that Mass could be said for him at Brentford and Ealing in 1505. This was a common pattern, not to say expected of a responsible citizen. Mrs Henry Yevele inherited all her husband's great wealth in 1400 on condition that she maintained two chaplains to celebrate Mass at the altar of St Mary in St Magnus' church for the souls of his first wife (Margaret), his parents (Roger and Marion), his siblings who had already died, King Edward III and Sir John Beauchamp, among others. Beauchamp was a patron and a friend who may have given Yevele that important first high-profile commission that brought him to the attention of other great patrons, leading to his becoming one of our greatest architects.

Having Masses said and praying for souls was how the medieval believer got their family and friends through Purgatory and into Paradise, and it was everyone's duty to do this. Some entered into business arrangements with the local monasteries and clergy so that they would not slip through the spiritual net. Richard Bischope for one, a carpenter by origin, got together with John Couper and the keepers of the Fraternity of the Resurrection in 1487 and came to an agreement with the prior of the Austin Friars of York that the convent would sing two trentals (sets of thirty) of Masses every year for the souls of the members of the Fraternity. Additional trentals of Masses were to be sung immediately a member died. Of course, they paid a kind of rent for this service as would only be reasonable, but it meant that they no longer had to worry about it since they had set up a kind of spiritual direct debit.

Reginald of Ely completely understood what was required to secure a place in Paradise. When he died in 1471 he made sure that he had paid off any 'tithes forgotten' at Bungay, Suffolk, where he had lived when he was working on the church of St Mary, and it seems that his wife died there. He left forty shillings for the building of a tower at Coltishall in Norfolk 'to pray for the souls of me and my parents'. Then he left twenty shillings for bridge building at Coltishall and 36s 8d for the same at Wroxham, plus 6s 8d for the Guild of the Holy Trinity at Sloley. Fifteen shillings also went to St Mary's, Bungay 'to pray for the souls of me and my wife'. He made financial arrangements for a chaplain fellow of Queen's College, Cambridge to say Mass for his soul daily for a year after his death as well as giving a piece of silver to a house of Trinitarian Friars at Thelsford in Warwickshire. Lastly, he ordered that rent from lands he had in

Barton and Comberton should be used to support three men in an almshouse he had founded. He was another person who asked to be buried before an image of St Christopher, this one being in St Botolph's church in Cambridge. It looks as though he had all the angles covered and, in fairness to other Master Masons who were not quite so bountiful, he had no children or even much family to provide for. Even so, a rough rule of thumb when writing your will in the Middle Ages was that you should give about a third of your assets to charity, whether specifically to the church or to a project useful to the community such as road building.[23] Simon of Canterbury managed this very neatly in 1341 because he had three tenements, one of which he earmarked for charity. He was another Master Mason who amassed great wealth and diverse property, including a brewery. He had begun his professional life as a carpenter and progressed to become one of the most important ones in London by the end of his career.

Tools being expensive, some Master Masons made sure that these ended up in the right place. Both Christopher Horner and one of the two Master Masons called John Herunden left their tools to be sold to raise money for the project they were working on at the time. In Herunden's case this included Twyford Bridge, so he effectively ticked the pious-giving box, too. Richard Aleyn stipulated that his tools should be sold in the will he made in late 1468 or early 1469. In addition, among other bequests to various churches, he left 20s for the repair of the roof at Kentford where, curiously he added, 'provided that the parishioners hold themselves thus contented, otherwise that they settle it as it shall seem best to them and my executors'.[24] This indicates that the work had not been completed and it sounds as though Aleyn's final illness came upon him fairly quickly, but he remained lucid enough to see that there might be problems finishing the assignment.

Sometimes the wills of the Middle Ages throw something to the surface that is not explained and is hard for us to fathom. The Master Mason Stephen Burton lived in the parish of St Michael upon Cornhill, London, where he had once been a church warden. His wife, Johane, had already died and he left the tools of his trade to be divided among his people, who were a mixture of friends and colleagues. He was a pious man and not wasteful; when he died in 1388 he requested that the four torches that he wanted to have used at his funeral should be used at other churches, so he did not expect them to be extravagantly burnt right down just for him. Good-quality candles were expensive and often feature in wills as part of the ceremony and as gifts, as we have seen in Chapter 2 with Sir Geoffrey Luttrell's will. He arranged for a trental of Masses to be said for his wife and for him and then he bequeathed four pence 'to every pore woman of my Crafte within London'. His craft (in the sense of his trade) was originally that of a mason and, as far as we know, there were no female masons at that time although women were by no means excluded from trades that might be thought to be masculine – there were lady butchers, blacksmiths and physicians, for example. The most likely explanation is that these women were widows or daughters of ordinary masons and stone-cutters or who were associated with a fraternity or guild to which he belonged since guilds were sometimes also known as 'crafts'. These must be women who had fallen on hard times on whom Burton had taken pity, but the wording of his will was clearer to him and to his executors than to us.

Simon the Mason's will of 1322 left to his wife Alice 'one acre of land in Boutham, at the rent of a rose in the season of roses'. He also ordered that a pig should be distributed (presumably in edible form) among those who attended his funeral. The rose is intriguing; it sounds romantic but it was more likely to have been the equivalent of a peppercorn rent in lieu of a service that would normally be given, or what was known as a 'quit-rent'. It is not clear how he came to bequeath that to his wife unless he wanted it written down as a reminder for whoever actually owned the land.

John Hampton's life may have been easier to unravel. In 1474 he was left 3s 4d by William Canynge, but only if he went to Canynge's funeral. As it is likely that Hampton is the same person who once had his pay docked in 1442 for getting to work late and leaving early, it may be that he was just a tiny bit lazy; his friend Canynge may have thought so too, but was going to get him to his funeral if it was the last thing he did – which it probably was.

CHAPTER FOUR

THE MASTER MASONS II

Builders behaving badly

WE KNOW SOMETHING about the lives of these great cathedral and church builders through their contracts – and sometimes through their failing to fulfil their
contracts, and by other legal documents. The Master Masons and their staff were
human and got into trouble with the law just as much as people will do in any age
and any society. There is even a case of a free fight breaking out in August 1324
between the monks and servants of Westminster and the masons; that particular
fight had the distinction of being recorded, and it is probable that there will
have been many more. That was a bad fight and ended in an ordinary mason
called Roger Alomaly being killed by a monk, Brother Robert de Kertlington.[1]
The cause of the fight is not known, but it was not so unusual for even a Master
Mason to be assaulted. The two worst cases concerned William Colchestre in
1408 and William East in 1512.

William Colchestre had been appointed by Henry IV to rebuild the belfry
at York Minster, which had fallen down. He had been described to the king
as a 'mason, expert in that art and much commended'. The local masons were
evidently less impressed than the king had been and, essentially, beat up both
William and his assistant, John Long, who would later become a Master Mason
in his own right. Although they suffered severe injuries, they obviously recovered
because they are both recorded as having run other building sites later. William
East very nearly lost his life when members of Brasenose Hall not only set upon
him, but threatened his colleague, Humphrey Coke, as well as William Vertue,
who was leading the building of Corpus Christi College, Oxford. Had anything
more sinister occurred, we could have lost three of our most distinguished medieval architects! The squabble went on for a couple of years until August 1514

when Formby (who had been a Principal of Brasenose) was bound over to keep the peace towards East and his colleagues and East, in his turn, undertook not to prosecute Formby outside the university. We assume that they wanted to keep the bad feeling in-house. The first of these cases was probably to do with local resentment at an outsider being brought in, but the second is less clear. Brasenose was a new college, built a few years before Corpus Christi, so it may have been merely a matter of jealousies between two new colleges.

The Ramsey family

William de Ramsey III, a member of the prolific Ramsey family of architects, complained in 1345 that his servants had been attacked and robbed by the abbot of St Augustine's, Canterbury (more likely by his staff than by the man himself). William de Helpestone was assaulted in 1375 by a gang who thought he was the William de Helpestone who was employed by the abbot of Royal Vale. It is not a common name so one might understand their mistake, although that was hardly an excuse for their anti-social behaviour. It is not known what grievance they had against their real target and the Helpestone who suffered the attack probably decided not to wait around long enough to discuss it with them.

The Ramseys were no strangers to the law and come across as having been a fairly ruthless family in their dealings. They were a clan of brilliant architects. It was William de Ramsey III who pioneered the distinctive Perpendicular style with its strong vertical lines and flattened-peak windows, but he was only one of seven members of this family listed as achieving Master Mason status within three generations between 1294 and 1371, operating mainly in London and Norwich. Seven in one family is not only a record, but also extraordinary, and without doubt the third William Ramsey was the most talented and prolific of the family. However, it was also he who instigated the kidnapping of a husband for his daughter, Agnes, then aged ten or eleven.[2] The candidate they chose was a fourteen-year-old called Robert Huberd, who was under the guardianship of one John Spray, but, more relevantly to the Ramseys, the lad was heir to a considerable fortune. William Ramsey III hatched the plot with his wife, Christina, his father William de Ramsey I and his uncle Nicholas. Although all of them were there on the day, William senior managed to keep his head down and was not actually brought before the court as his son and daughter-in-law were a few days after the incident.

All that we know of the actual event was that on 20 November 1331 William and Christina de Ramsey and some others who included his family members, a man named Thomas de Chacombe, and a priest called John of Wolcherche went to John Spray's house outside Aldersgate, forcibly removed the teenager and compelled him to marry Agnes. Her parents and Chacombe were arrested and charged but, since the marriage had happened, that at least was a *fait accompli*. The court proceedings also said that it could not be annulled, which we must interpret as we wish, but it should be noted that a marriage at that time was not merely a matter of exchanging vows in front of a priest at the church door (not inside the church as we do now), but involved previous agreements or contracts, consent by both parties and consummation.[3] The family pleaded not guilty to the abduction and claimed that the two youngsters had wanted to marry, which perhaps they had, although that was beside the point. The court asked for an account of all Robert Huberd's property and ordered that he should stay with the Chamberlain while the matter was looked into. In the end, the groom was asked where he wanted to live and he chose the Ramseys with his new bride, so it is to be hoped that they lived happily ever after. Even without the abduction scandal, the case may have been a local *cause célèbre* because under-age marriage was not as common as might be thought. At the time that the Ramseys set off for Aldersgate with their young daughter, the general age for marriage among non-aristocracy was early twenties in the country and mid-twenties in towns.[4] Even given that there was not yet a legally fixed earliest age for marriage, Agnes was exceptionally young since maturity was based on puberty. The age at which a girl was considered old enough to make mature decisions about things like marriage, working away from home, entering a nunnery or similar was twelve, whereas it was fourteen for a boy.[5] Society was shocked, but probably more because the parents had effectively sold their daughter for cash than because of her young age. It shows the Ramseys in an unsavoury light, not least because they did not need the money, being prosperous in their own right. By a twist of fate, William de Ramsey III died from the Black Death in 1349 and all those involved in this distasteful episode probably did as well because the date of death for a number of the Ramseys is also 1349. That would have chimed with medieval moralizing: no good can come to the bad (sadly untrue).

Murder, money woes and other anti-social behaviour

The Ramseys were far from alone in being hauled up in front of the courts for unlawful behaviour. Although they produced some of our greatest and most tranquil spiritual buildings, they too were mere mortals. Some became distracted from spiritual things, not least by spending time in prison, as did William Hyndeley. He had designed the rood-screen in York Minster sometime after 1475 and had been Warden of the York Minster Masons' lodge for nearly six months in 1472, and so was a man of standing in his community. He was sent to jail in 1490 while an investigation was held to see if he and another Master Mason, Christopher Horner, had murdered a tiler called John Partrik. This had happened during a dispute between the masons and tilers; these types of dispute were not at all uncommon, were usually about pay and conditions, and were sometimes long-running. They must have been the bane of Master Masons' lives, operating as they had to within a budget and to a deadline. The two men must have been found not guilty because they are both heard of again later (this still being a time when murder was a capital offence) and Hyndeley did not die until 1505. His alleged partner-in-crime, Horner, was of a more belligerent character altogether. He succeeded Hyndeley as Master Mason at York Minster, but was thrown into jail again in 1504 for causing a disturbance at the election of the mayor. He was incarcerated that time for about five months and then jailed again later for similar behaviour. That said, he lived until 1523, when he would have been in his fifties and no doubt sustained by his long-suffering wife, Agnes, whom he married in 1499.

One long-suffering husband was William Wetyng who married a widow of property from Peterborough. He himself was working at Westminster when he was sent up to that area to press men for the work and took his bride with him. Perhaps she wanted to visit old friends and show off her new husband. This turned out to be a mistake. When they reached the city, both were immediately arrested and jailed by the abbot because of unpaid rents, which the new Mrs Wetyng should have dealt with when she was acting as the executrix of her former husband's will. He had been the abbot's bailiff and therefore a man of some status. This is unlikely to have found favour with William Wetyng and it seems naive of Mrs Wetyng; as wife of the bailiff she must have been aware of the laws and administration surrounding wills; her previous husband must have thought her competent or he would not have appointed her as executrix. A bailiff at this time, it should be mentioned, was not confiscating goods to off-set a bad debt, but was an officer representing the monarch, or a justice officer working for the sheriff, or the steward of the lord of the manor who would collect rents. The word has shifted in its meaning over the years; in 1506 when the Wetyngs were arrested, the bailiff's appointment commanded respect.

In other cases, we see Simon Pabenham and Richard de Wytham having such a row in 1298 that they were eventually called in front of the mayor and aldermen at the Guildhall, London, where they were reconciled in the sense that they were made to agree that the first one to renew the quarrel had to give five pounds (a considerable sum of money then) towards the building of London Bridge. This was an excellent way of stopping any argument and one envisages a certain amount of biting of tongues. This makes it unlikely to have been a private row, but more the sort that involved most of the workforce and invited others to take sides making progress on the building site impracticable. In 1439 William Layer, then working at Bury St Edmunds, was summoned along with twenty-nine others for throwing garbage into the streets near the abbey, but this seems to have been more out of Bury's desire to clean up the streets than any ill-temper on the part of the people. That cannot be said of Robert Kentbury, working with William Wynford at Westminster in 1397, when he so lost his temper that on 28 June of that year he was arrested for treading down and wasting the abbot of Westminster's crops. It is easy to imagine him jumping up and down, incandescent with rage; perhaps the abbot had offered some need-lessly helpful advice of the 'you've-missed-a-bit' variety and it shows that this Master Mason at least saw himself as an artiste. Some behaved in a somewhat high-handed manner abducting, so to speak, a cartload of bread as Richard Lenginour did in 1290. He argued successfully that this cartload was being taken to Chester by someone who was not local and so this man's business was damaging to trade already in place, so he was only safeguarding the dealings of the Dee Mills (in which he had an interest, let it be said quietly).

John Hassock and Lawrence Holbrook, both from Rochester, were par-doned for their part in Jack Cade's rebellion, an uprising by men from Kent who were small landowners rather than members of the aristocracy. These rebels objected to much of Henry VI's government under which they felt they suffered forced labour, corrupt courts, the wrongful seizing of their lands, the damag-ing effect of the loss of France, and heavy taxation. It is not entirely clear who Jack Cade was, or even what his real name was, but under his leadership a mob gathered in Kent where they managed to overcome a government force sent to disperse them, and then marched on London. Initially, they had the backing of Londoners, but they overstepped the mark in violent and yobbish behaviour that turned the City against them. Most of the mob accepted a pardon issued by the king and went home, so Hassock and Holbrook must have been among this number. Neither man became ranked among the highest of the Master Masons, but that reflected their ability more than their politics, although others fared less well when the fortunes of the country changed. Simon Clenchwarton and others lost their positions in 1461 on the accession of Edward IV. He and other King's Master Carpenters and King's Master Masons found themselves suspected of being loyal to the House of Lancaster and thus being out on their ears to the benefit of others, such as Edmund Graveley, who appeared in the accounts as having taken over Clenchwarton's work. Clenchwarton was given a general pardon, but not his job back; the fact that Graveley was paid eight pence a day less than he had been, which was a significant amount, may not have been much consolation to him.

As we go through these cases, we can see that little changes in human nature. It is easy to gaze in wonder and marvel at our great cathedral buildings but, given the dispositions of some of the architects, perhaps we should wonder and marvel that any of them were made at all. Some of our greatest medieval designers were all but impossible to work with.

Absentee builders

Christopher Scune was appointed to carry on with the work which John Cole had started, building a steeple for the church at Louth in Lincolnshire in 1505. He, however, was already working on the rebuilding of the nave at Ripon cathedral, a much more prestigious venture. This work would go on until about 1515 so the Louth steeple did not stand much chance of getting his full attention, and one could say that he should not have taken on the project in the first place. He was away from the work at Louth for more than six weeks at a time and then visited increasingly rarely in July 1506, August 1507, April 1508 and 1509, and so on, until things came to a head in 1515. Letters had been sent to him asking for his presence at the building site, but these do not seem to have had the desired effect. In the end, a man called Lawrence Mason was paid to ride to 'his master in the north'. Lawrence must have been the Under-Master Mason or leader of the masons working at Louth or, at any rate, someone very familiar with details of the work. Lawrence does not seem to have been successful since he reported back that Scune had said that he would have no more dealings with the work, although he does seem to have given some advice and charged a consultancy fee of 6s 8d. Although that seems a bit of a cheek, it may have been welcome news because it ended a ten-year, difficult working relationship and cleared the way for the Works Committee of Louth church to engage the services of John Tempas, who finished the job. It was near completion, anyway, and Tempas achieved this between 25 June and 15 September 1515, although the committee must have felt their hearts sinking when they found that Tempas, too, had disappeared at the beginning of August and must have wondered if history was about to repeat itself. Fortunately, Tempas was only away for a week and then for another week later on, and the job was done. It is thought that it was the same John Tempas who designed the lacy, octagonal lantern on top of the tower at St Botolph's church, Boston, which can be seen for miles. The walls of this lantern are so thin that there is no question of any further structure. The whole tower was probably started by Reginald of Ely in the middle of the fifteenth century. The fact that it stands so solidly today is a tribute to its foundations because the Master Mason dug down to the firm boulder clay, which is some 5ft deeper than the bed of the river Haven.

Unfortunately, no records survive of how this enormous pit was achieved, other than by plain digging by squads of men, but there must have been a way to keep it drained as they dug.

The better Master Masons were continually approached, poached and tempted by bigger and better commissions. We are all familiar with the builder who assesses the job, is either pessimistic or over-enthusiastic, works for a week and then disappears leaving a mess and things hardly begun, and there is nothing new in that. William Lyngwode from Blofield, near Norwich, was creating the choir stalls in Winchester cathedral in 1308 when he was required back in Norfolk to do his annual duty at the manor court. The bishop of Winchester wrote to the bishop of Norwich asking if Lyngwode could carry on with the stalls and have a stay of execution until Michaelmas (29 September) 1309, but the job evidently was not finished by then because a second letter was written asking for a dispensation indefinitely. What is clear is that the bishop of Winchester strongly suspected that if he let Lyngwode go, he would be lucky if he ever saw him again and his precious stalls would be left half-finished forever. The request was granted and visitors to Winchester cathedral today can appreciate the bishop of Norwich's forbearance in this matter; the choir stalls are a joy and include a set of sixty-eight misericords that have the distinction of being the earliest set of carvings in our cathedrals and churches to show animals as domestic pets as opposed to working ones or monsters. Misericords are the tip-up, theatre-style seats found in over 700 churches in England, and numerous ones elsewhere in Britain and overseas. The idea was that the monks and clergy could ease the misery of their aching joints during the long services by perching on them when they were in the up position (misericord comes from the Latin for 'mercy'). If they rested too heavily, the seat would crash down into the seated position, certainly drawing attention to the monk, if not actually sending him sprawling. On the underside of these seats there are frequently carvings, the majority of which have nothing to do with religion; some are funny or comments on daily life, others are grotesque or downright bawdy. As a form of medieval art, they have survived very well because they could be placed face-down and therefore be out of harm's way. Also, because so few of the carvings are overtly religious they usually escaped mutilation in various times of spiritual enlightenment.

Enforcing a contract with a builder

The common story of Master Masons starting a job, getting a better offer and moving on to the next project without finishing the first seems to have been a problem not just throughout the ages, but across Europe; there are numerous contracts that were worded to try to prevent this from happening. In 1252, the man chosen to be the Master Mason for Meaux cathedral had to agree:

> In addition, he cannot accept any work outside the diocese without our permission … He will not have the right to go to the site at Evreux or to any site outside Meaux, or to remain there longer than two months, without the permission of the Chapter of Meaux. He will be obliged to live in the town of Meaux and he has sworn that he will work faithfully on the above-named building site and will remain loyal to it.[6]

Unfortunately, we do not know the details of any contract he had made with the Works Committee at Evreux, or if they opposed the commitment to Meaux. It is possible that there was no such contract at that stage; the Meaux Works Committee could well have been trying to cover every loophole as they would certainly have known if building works were being discussed at Evreux. They would also have been keenly aware that good Master Masons were in danger of being poached.

Others were rather more stringent, and the following extract from a contract made at Troyes for roofing the cathedral there on 11 October 1390 is redolent with the building-site experience of the man who drew it up – or perhaps he had worked with the Nepveu brothers before and was anticipating trouble? Having stipulated from where Jehan and Colart Nepveu were to obtain building materials and the precise dimensions of the roofing area to be worked, the contract then detailed the money that they would receive and how they would be paid in instalments – then came the sting. The brothers were required to give their word that they would finish the work:

> Under pain of arrest, of being put and kept in prison, and against a guarantee of all their goods and the goods of their heirs, movables and real estate, now and to come, submitting and giving as guarantee the said goods to the judgement and decree of our Lord the King, of his men and of all other men of justice, the said brothers have undertaken … to make, accomplish perfectly, finish and successfully conclude all the above-named things in the manner stated above, without any shortcoming, without deviation, on pain of having to reimburse and pay in return all costs and damages which might arise or depend on this.[7]

The Nepveus then had to agree that they would have no right to appeal if things did not work out as they intended. This type of contract is strongly worded, but not so unusual: the Master Mason who took on building the nave at Fotheringay,

Northamptonshire, in 1434 was made aware that if he did not complete the work 'within a reasonable time to be agreed upon', he could also find himself in prison.

William Reyner, a carpenter by origin, seems to have been one of our most elusive Master Masons. He was a Suffolk man who was contracted on 8 June 1413 to make the woodwork for a new bakery, which was to be built at King's Hall, Cambridge (now Trinity College). As ever, things seem to have begun well and Reyner appears to have begun amassing the appropriate materials. It was not long before things started to go astray, however, because the accounts fairly soon show that payments were made to a man called Watton who had a horse and who could ride for two days to find out where Reyner and his bondsman, Robert Strut, had gone to. It is not clear whether Strut was a craftsman or was involved in the financial side of the business; either way, he was nowhere to be found when the Master Mason was needed. A Richard Wrythe was then paid to go to Mildenhall, where it must have been thought that Reyner had taken on another job. In the end, things became very strained with writs of *Capias* and *Exigent* being taken out against him. A *Capias* was a writ allowing for Reyner to be arrested and an *Exigent* was a writ commanding the sheriff to summon the defendant to appear on pain of outlawry. This is about as bad as it could get without the Master Mason just plain refusing to go on as Christopher Scune had done. There was evidently a complete breakdown of any working relationship.

Working relationships

Sometimes things were arranged in an orderly fashion: Oliver de Stainefield is likely to have been the Master Mason who designed the nave at Beverley Minster in about 1308. He was, in fact, given permission to leave this work when the Earl of Lincoln requested his services elsewhere. So that Stainefield did not get too carried away with his new tasks, it was agreed that he should pay a mark (13s 4d) a year to Beverley's building funds until he returned. This was designed to concentrate his mind so that he did not forget his obligation to Beverley. Of course, it must be said that the majority of Master Masons were organized and honest men. John Lewyn may be typical, a man who was the principal mason to Durham cathedral in the latter part of the fourteenth century and is heard of steadily from about 1364 to 1398 working for ecclesiastical and royal clients. He ran several concurrent high-profile projects in the north without apparent difficulty. He specialized in building for strength and so was in demand whenever there was anxiety about raids from Scotland, but his own strength was his adaptability, not just in design but in business. He was not just an architect, but also a businessman who exported wool on a large scale.[8]

When working relationships failed

All was not necessarily sweetness and light when the Master Mason was present on site. In April 1398, Robert Scot petitioned Richard II that he should be given the job of Master Carpenter for the county of Chester in place of William de Newhall who, Scot said, was too old and feeble to do the job. He did not get his way for twenty-one months when he finally got the appointment he craved, only to have it taken from him a few months later and given back to Newhall for the rest of his life (he lived for another decade). Out of these lines, we can pick a few things, one of which is that Scot exaggerated how old and feeble Newhall was (further borne out by the fact that there is a record of Newhall having an apprentice in 1401/2). It also shows how keen the competition was for these royal appointments and how desirable they were. The action is unlikely to have improved relations between the two men.

Rows and arguments seem to have been ongoing in the building of our churches, but it was not always the Master Mason who was at fault. A case is recorded in England at Spalding, in Lincolnshire, at the Visitation. The Visitation is a periodic formal inspection of temporal and spiritual affairs of a diocese under a bishop's control and would involve specifically visiting each location. On 1 October 1439, the inspectors of the Benedictine Priory of Spalding were told that one of the monks, Brother John Bostone the elder, had been so much of a curmudgeon that the ordinary masons had left and that the Master Mason would only stay if he were guaranteed a life pension from the monastery. This was on top of his already having been offered a very respectable fee of twelve marks (eight pounds) a year. One cannot help but wonder if that particular Master Mason was hoping to exploit an already difficult situation.

So how did they get away with this sort of behaviour? While those records that survive are more likely to show their excellent work and praise the Master Masons rather than criticize them, naturally they did have critics. Nicolas de Biard was one such. He was a Dominican preacher in Paris in the second half of the thirteenth century and was probably one of those people who regard anyone in management as idle and are not able to see any benefit in having one person who plans, coordinates and takes the responsibility. In some of his sermons he said, for example:

The Master Masons, holding measuring rod and gloves in their hands, say to the others: 'Cut here' and they do not work; nevertheless they receive the greater fees, as do many modern churchmen. Some work with words only. Observe: in these large buildings there is wont to be one chief master who orders matters only by word, rarely or never putting his hand to the task, but nevertheless receiving higher wages than the others. So there are many in the church who have rich benefices, and God knows how much good they do; they work with the tongue alone, saying, 'Thus should you do,' and they themselves do nothing.[9]

Considering how much building was going on just in the area we now call France and how much employment it gave, the sermon had little impact: the ordinary workman did not get employment without these allegedly exploitative Master Masons.

'A most learned master of ... masonry'

Mainly, however, the Master Masons were praised, and the records ring richly with appreciation of their work. Blithere, a Saxon and Master of the Works at Canterbury cathedral in 1090 was described as being, 'the very distinguished master of the craftsmen and director of the beautiful church'.[10] In about 1100, Robert the Mason was said to have surpassed all the masons of his time at St Albans. Around 1150, Hugh the Carver, who made the doors for the abbey at Bury St Edmunds, drew the comment: 'as he outshone all men in his other works, in this marvellous piece of work excelled himself'.[11] In 1113, Arnold the Mason, a lay brother of Croyland (or Crowland) Abbey, in Lincolnshire, was reported as being 'a most learned master of the art of masonry'.[12] Robert Man, a Benedictine monk at Peterborough from 1425–c.1444, was described as being 'of no account in matters temporal ... albeit he has some degree of experience in the craft of stonemason and carpenter'.[13] This does not have quite the warmth of other plaudits, but Robert Man was probably pleased to have it said of him – but let us hope not too much as that would have contravened Rule 57 of the Rule of St Benedict ('If there are craftsmen in the monastery, let them carry on their crafts in all humility').[14]

Monks as Master Masons

As will be mentioned in Chapter 4, there were a handful of Master Masons who were also religious, but perhaps not as many as are popularly supposed. The fact that a guide book might say that a cathedral was built by a particular bishop, does not mean that that same holy gentleman ever picked up a shovel; it is more likely to mean that he was the instigator, or even the person who commissioned the work. With the exception of Peter of Colechurch, the priest whom we met in

Chapter 3, the remaining churchmen listed as being Master Masons were either monks or clerics. It is thought that a monk, Jean Marre, designed the cathedrals of Auch and Condom in the south of France at the end of the fifteenth century. The clerics can be a nebulous group since, although technically if a cleric this makes that person a member of the clergy, in this context, it does not mean that the individual concerned had ever taken Holy Orders. A number of young men were educated and trained by the Church with a view to them becoming administrators and some of these moved into architectural circles. Elias of Dereham was possibly the most famous of these: a highly influential figure who was present at the sealing of the Magna Carta in 1215 and, indeed, was one of the commissioners employed to distribute it. He was very much involved in the Translation of Thomas Becket's bones on 7 July 1220, together with Walter of Colchester, sacrist of St Albans, when the two of them were described as being incomparable in having done 'everything necessary to the making of the shrine, to its setting up and removal … without cause for blame'.[15] He was also in charge of the works of Salisbury cathedral, among numerous other projects, but more in the sense of being a project manager than architect.

Some monks outdid all others at their craft, which must have caused eyebrows to be raised, if not actual problems for them. The mention of Rule 57 of the Benedictines above is not intended to be flippant: a man entering a monastic order had to undergo stringent tests to see if he was suitable for what this entailed; did he have the necessary staying power, or was he just looking for a meal ticket for life? The Rules of St Augustine and especially that of St Benedict make this very plain. A monk was also required to attend half a dozen services a day, not all of them conveniently timed if someone was also trying to operate in what we would call the real world. In addition, he had to have permission from his abbot simply to leave the monastery. A monk must not gain personally from his secular work, and he must never be guilty of the sin of Ananias and Saphira, who sold a possession to make an offering, but kept back part of the price for themselves.[16] Cistercian monasteries usually produced their own builders, but sometimes found that they had no one with the right experience, as once happened at Buckfast Abbey, in Devon. On this occasion, a young monk was sent to France for eighteen months to study the craft of building, which is not long for such an important training, even if he was required only to construct the building, not design it as well. He was then tasked to do on-the-job training for five lay brothers, who were permitted to be absent from all the holy offices except for the first and last of the day so that work could proceed without too much interruption. Nonetheless, it took the monks at Buckfast about thirty-two years to finish the job.[17]

Some of these men worked in fields that were very different from their calling, such as Robert de Holmcultran, a Cistercian (albeit a lay brother) who worked on engines of war for Edward I, and Ralph of Northampton, an Augustinian canon who repaired Henry III's fishpond at Woodstock, in Oxfordshire. In 1333, Brother William of the House of St Robert of Knaresborough earned five pence a day for his work as a carver. One of the most outstanding of the religious Master Masons was Alan of Walsingham, a junior monk of Ely, who was originally noted for his work as a goldsmith. Even when otherwise occupied as sacrist, he took charge of the rebuilding at Ely, which included the remarkable octagon designed by William Hurley for which the present cathedral is so famous. Not all were as successful: the monk Thomas de Northwich was both a builder and a physician and built the tower of the abbey church of Northwich, in Cheshire, which collapsed soon after his death in 1207. The luckless Gilbert de Eversolt was a Benedictine monk who was put in charge of works at St Albans in about 1200 – alas, these simply did not progress under his direction, so it seems likely that he had been given the task when he had no real knowledge of how to set about it. He was indeed the 'Brother faced with the Impossible Task' who features in Chapter 68 of the Rule of St Benedict. On a happier note, William Corvehill, a monk of Wenlock Abbey from about 1500–46, was a useful man to have in any monastery because he was described as having been:

> Excellently and singularly expert in diverse of the seven liberal sciences and especially in geometry, not greatly by speculation, but by experience; and few or none of handy craft but that he had a very good insight in them, as the making of organs, of a clock and chimes, and in carving, in masonry, and weaving of silk, and in painting ... a good bell founder and a maker of the frame for bells.[18]

How were Master Masons regarded by medieval society?

The Master Masons may have been difficult to work with occasionally, but not only did towns, cities, the Church and the ordinary people want their end-product, but Master Masons were also personally held in extremely high esteem, which was the main reason why occasional prima donna-ish behaviour was tolerated. We still look at their works in wonder. They were held in such regard that they might even attend the funeral of a monarch as John Massingham, William de Wynford and Henry Yevele did when Queen Philippa, the wife of Edward III, died in 1369. They were issued with cloth to make mourning clothes for the occasion. They were regularly invited to join the high table of colleges in Oxford or Cambridge or that of a local lord. Thomas Sturgeon dined formally at the high table of King's College no fewer than fourteen times between March 1459 and

April 1461 when he was working on the chapel. He was once accompanied by his father, who no doubt was extremely proud. In the 1390s William Brown similarly dined with the Fellows of New College Oxford despite his having been fined four times under the Statute of Labourers for taking excessive wages. However, this may not have been seen as being a particularly heinous crime, rather one that was more of an administrative error as the Statute existed to try to keep some sort of control, especially after plague had reduced the workforce.

Enjoying the perks of the job

The main way to assess the status of Master Masons is to consider the perks of their job and the very great wealth some amassed, as well as the fact that a few also achieved recognition from the town by being appointed to some of the more prestigious civic posts. This was all the more remarkable because those connected with the building trades, being more often itinerant in nature, were not normally those chosen to lead the community, as will be discussed in Chapter 7. That said, Thomas le Tighelere had considerable success. He lived in Farnham, in Surrey, and seems mainly to have been based there, so he knew and understood the local people and their issues. He was a multi-tasker in that he was a carpenter, a miller and a contractor and seems to have been able to turn his hand to most types of work ranging from repairing the castle roofs, building houses, making dykes and sluices for the pond at Bourne Mill and also for the bishop's mill at Southwark, to repairing the head of Frensham Pond, and so on. In 1309, he represented Farnham in Parliament and so was one of the town's first two MPs. This was not an office granted to anyone who wished to apply to canvass for votes as it is today, but shows that he was a man of very great prestige. Only three other Master Masons are recorded as having reached that rank. One of these was John of Bury, who became MP for Cambridge in 1514, having also been bailiff in 1502, 1503 and 1506, and mayor in 1510 and 1517 (not bad for a man who is thought to have had a stutter). He died in 1522, possibly from trying to juggle too many jobs at once! A mason by trade, John of Bury seems to have run a successful bricks and tiles business as well as manage numerous building projects. An even greater collector of titles was Thomas Drawswerd from York, who was an imager and carver. In 1511/12 he became MP for York, but his civic career seems to have begun in 1501 when he was Chamberlain of York and then Sheriff in 1506–7. The following year he was elected alderman in York. At that time, he was working on the great roodscreen for Newark church, in Nottinghamshire, so a lot of his time must have been spent riding between the two towns. Having been MP, he was chosen to be Lord Mayor of York in 1515 and again in 1523. These appointments came at no small expense to the individual as Drawswerd found since the records show that he had spent £22 on civic offices over the years. A third Master Mason, Richard Russell, became an MP in 1420 and a bailiff of Dunwich in 1430 and 1440 while at the same time designing Kessingland church, but it was more likely that a Master Mason (were he to be appointed to a civic office) would be an alderman as were John Marwe in 1442, John Forster in 1473, John Kilham in 1487 and Nicholas Reuell in 1520. An alderman might have been the senior member of a guild, or he might have held an office in a town or district working in support of the mayor, so these men were very much committed to their communities in terms of spending time away from their main work so that they could be considered for election. Finding people to do this type of work was not always easy and a man might be required to be a jurat (one of twelve men who, in effect, worked with the mayor and bailiff and might now be called the town council), and might have to pay a fine if he could not fulfil the role. It should not be thought that town officials were automatically held in high respect by the common people, despite the prestige of the actual appointment. In Sandwich, in Kent, the town records show that in 1529 a Robert Kenny was accused of calling the mayor and jurats 'hedgehogges, hedgecreepers, benche whystlers and catchpollers' and, shockingly, had declared that if he met the mayor in the street he would not doff his cap to him.[19]

Not all these Master Masons were necessarily cut out for civic duty simply because they were appointed. Alexander Tyneham was Provost of Yeovil, in Somerset, in September 1387, but was asked to stand down. It seems that he had no intention of giving up the perks that went with the job and eventually the burgesses of the town had to petition the chancellor to make Tyneham surrender the town seal and other insignia after he had been dismissed. It sounds as if he was certainly at odds with his staff, although the record does not reveal what had triggered his removal from office. The seal is still in use so Tyneham must have handed it in, but how graciously (or otherwise) we can only guess.

Such was the prestige of the position of the medieval architect that it was not uncommon for him to be granted the freedom of the city or borough, although many would already have been freemen or burgesses either through birth or marriage. For instance, William Atwood married the widow of a freeman, which gave him free status in 1498. This was also something that could be applied for, with an entrance fee levied. In fifteenth-century Romney, in Kent, the freedom fine ranged from 2s to 10s and almost 400 freemen were enrolled from a variety of backgrounds – albeit that one criteria given was 'if a stranger, he be of good name and conversation'.[20] Freemen had quite a few privileges so freedom of a town or city was distinctly worth having. In Canterbury c.1430, there were nearly thirty clauses, which was not untypical, including:

That the freemen may come to the council of the city and there speak and be heard, whereas others must keep away or be put out....

Also freemen of the city may exercise a craft and open windows without leave [open a shop which consisted of using a ground-floor room with a window opening onto the street], whereas others must make agreement and come to terms with the chamber [city treasurer] of the said city....

Also freemen of the city are quit of toll customs, of lastage, and of shewing throughout England, and along sea-coasts, as allowed by their charters, whereas others do not have this privilege.[21]

Lastage was a tax paid on goods and merchandise, such as wool or grain, which were sold in units of 'last' – quite a lucrative perk. Nor were the perks limited to economic ones; often there were legal rights such as the right to hunt and not pay *Earesgeue*, which is thought to have been a New Year's gift to the sheriff.

These privileges were worth having and could be taken away if a freeman transgressed. The Calendar Letter-Books of 1370 record a mercer named Richard de Northbury, who had married Imanya, the widow and executrix of John de Enefeld, a pepperer. Between them they had failed to execute a part of Enefield's will, which they had previously been ordered to do by the mayor. Consequently, 'Richard being a freeman of the City, and having appealed to the court Christian, contrary to the liberties of the City, is ordered to lose his freedom.' This will have affected Richard's trade and his reputation badly, and probably led to a certain amount of domestic strife as well. Four years later, he regained his freedom when he appeared before the mayor, and the aldermen and paid a fine; further proof, were it needed, that freedom in the legal and administrative sense was valuable.

Although the Master Masons were men of substance, individual builders (with all the trades that this term implies) were not. In fact, basic masons were often the poor men of the city. Of all the ordinary masons who were made free in York in the fifteenth century, only seven per cent of them had enough assets to make it worthwhile making a will, as compared to twenty per cent of tanners.[22]

The Master Masons were expected to be involved in the administration of the community by taking their turn on things like jury service just as much as other men of status, and this is evident from the number of times they tried to get out of it. It would be unfair to say that in 1281 Richard Crundale was one of the few to carry out this important duty, but the records seem to report a greater interest in avoiding it than in doing it, and one can see why works committees were eager to support their petitions; it tied in with contracts insisting that Master Masons did not travel far from the building site. The works committees would probably have gone along with almost anything just to keep the Master Mason focused and in attendance, although exemption was also linked to reward. In 1228, John of Gloucester was not only excused from jury service and 'certain taxes', but was also granted a house in Bridport, Dorset, as well as two more in Oxford for his good service to King Henry III. He had already been given an estate at Bletchingdon, Oxfordshire, land in Middlesex and Surrey, and some property in Southwark. He was apparently on good terms with the king at a personal level as there is an intriguing mention of Henry III returning five casks of wine to him to replace ones taken at Oxford.

Thomas Hoo was not the main architect for the rebuilding of the nave at Canterbury cathedral in 1380 (he was working for Henry Yevele). Nonetheless, he was allowed two years' exemption from jury service in 1380 and a further three years from 1389/90. Yevele was working on various other projects at the time in the City of London and elsewhere; a Master Mason of his stature would have delegated to a talented and competent architect such as Hoo once the plan had been decided and the work set in motion. John Lewyn was another Master Mason of considerable renown, working in the north and being principal mason to Durham cathedral, among other appointments. As Bishop's Mason, he was excused jury service in 1368–9 and was also paid handsomely throughout his career and given lands. William de Ramsey III might have felt that he had had enough to do with juries through his previous escapades, but so that he could give his whole and undivided attention to the design and building of the chapter house and cloisters at St Paul's, London, he was excused his civic duty in 1332. However, Stephen Lote won the 'no-jury' service stakes since he was made exempt for life on 5 December 1414, but by then he was an elderly man, dying three or four years later. The entry in the Calendar of Patent Rolls also gives him the right that:

> none of the king's lieges or any alien shall be lodged in his house, buildings
> or other possessions and nothing taken of his corn, horses, carts, carriages or
> victuals or other goods against his will.[23]

That he was given that privilege is interesting, but even more so is the fact that at this time the conditions existed allowing such a privilege to be given. These small entries reveal how everyday life was in the fourteenth century and that people with land, for example, had to petition or hope to be granted freedom from having their crops annexed, or being used as a bed and breakfast for passing royal officials.

Other perks might be applied for if these were not freely given. Exemption from paying some taxes has been mentioned and that appears to have been arranged on a very local basis. In 1370, four Master Masons joined together to get the right to be exempt from paying taxes and subsidies in London on the grounds that they were then the four sworn-masons and sworn-carpenters (in effect, surveyors) of the city, and they had discovered that their predecessors doing that job had been so exempt for the last century. This was approved for as long as they stayed in office. On the other side of the coin, a hazard of the job was that a Master Mason could be pressed for service in much the same way that in the eighteenth and early nineteenth centuries men could be press-ganged into the Royal Navy. Master Masons were issued with licences to do the same to get men for the king's works. This sounds draconian, but it may be that the men involved were not too concerned because it meant a guarantee of employment at least for a few weeks. One man was able to turn this to his advantage when he was sued for breach of contract in 1530: John Hawkins was able to defend himself against Katherine Adams' claim that he had failed to build two houses for her in Shoreditch because he and all his servants had been pressed to work at York Palace. To save any such annoyance, it was better to get exemption from impressment in the first place, which was the route taken by Thomas Denyar in 1414 and allowed for the duration of the building of St John the Baptist's chapel at Hereford cathedral.

Some perks were seen as such, but were, in fact, part and parcel of the way negotiations were done and contracts drawn up. The Master Masons were usually extremely well paid, but it was not always in cash or even in items. As they progressed in their profession and gained more and more prestige, it became more likely that they would be appointed for life, or to have themselves and their families taken care of right up until their deaths (and often beyond in that their widows were sometimes provided for). This would have been an important part of the bargaining since these are the days when there was no safety net of the Welfare State available for the poor, infirm or elderly. If a Master Mason had not made enough to support himself and his family at the end of his life, he would have to rely on the charity of his neighbours, or starve and hope someone would give him a decent burial.

Examples of Master Masons being appointed for life are numerous, but many of the contracts were qualified for fear of the individual living to a great age, as sometimes might happen. Richard Lenginour had a straightforward grant of 12d a day for life in 1284 and did well out of the bargain – he survived for about another three decades and was probably in his seventies when he died. However, he does not appear to have had any surviving family and would have had to arrange his own care if he needed it. In 1278, Walter of Hereford had a much more detailed arrangement with the abbot of Winchcombe. This contained the now familiar caveat that, 'Other work, except the works of our Lord the King, he shall not begin without leave of the Abbot and Convent', and later it listed his entitlement if he was ever ill enough to have to stay in his room: 'two monastic loaves daily, two jugs of convent ale and two dishes from the abbot's kitchen, and for his two servants such allowance as the abbot's servants receive'. If he ended up living with them forever because of infirmity then the entitlement would drop to an allowance for only one servant and one horse. He would also have his annual robe of office and be allowed to build himself a chamber within the grounds of the monastery. He was allowed two wax candles every night, four tallow ones and a load of firewood every week from Michaelmas to Easter.[24] This seems to have been a generous provision, but it is by no means unusual, occurring all the way through the Middle Ages. At the other end of the period, John Courtley was given a job for life at Quarr Abbey, on the Isle of Wight. His contract gave him 26s 8d a year, with a robe of the livery of the abbot's yeomen. He was also granted a house and some land close to the monastery, a ration of two loaves of convent bread, two loaves of household bread, three gallons of convent ale. He was given as many meals of flesh or fish as a Quarr monk would have, as well as an allowance of firewood, with a similar allocation made to Joan, Courtley's wife. This was a weekly allowance lest anyone should be concerned about the Courtleys' livers, and it was to continue until both the Courtleys were dead.

In 1471, Thomas Peyntour was granted a patent for life by Ely cathedral. This was to bring him eleven marks a year (about £6), a suit of clothes and a fortnight's holiday. When he became too old to work, the money was to be reduced to six marks (about £2.50) and he would also have his livery. In 1488, the prior and Chapter of Durham gave John Bell junior a house and a pension of four marks (£2 13s 4d) a year, having previously stipulated that he must not depart from his occupation without a special licence, so he was not allowed to make any money on the side from other areas of work. Lincoln cathedral allocated Robert de Bokinghale a yearly pension of nine marks in 1307 for as long as he was able to be the chief carpenter, and this was to be for his lifetime so long as he stuck to the letter of his contract (which is likely to have mentioned that he must not stray elsewhere).

Things did not always work out as perhaps had been intended. Hugh Herland, renowned for making the hammer-beam roof of Westminster Hall (among numerous other works), was granted equally plentiful benefits to reflect his great skill. However, when he was quite an old man he had to petition to get a particular pension of ten marks a year at Winchester, which does not appear to have been paid to him for some four years. In his case, he had amassed enough riches not to have a pressing need of the pension, which also often happened. Quite a few executors chased up monies owed to Master Masons, suggesting that they had more than enough to live off while alive.

Some may have been a little profligate with money. The case of Thomas Stafford in 1512 is unusual for a Master Mason in that he was given a number of benefits at Arbury that would continue up until his death. However, when he was no longer able to work, it was specified that he would have a house, with food and drink and even access to a barber and laundry as well as fuel and pasturing for a horse or a cow, but there was to be no money attached to the deal. The most probable explanation is that Arbury Priory, in Warwickshire, did not want to be locked into cash payments that could become difficult, whereas the priory may have felt that it would always be able to provide the other items. It is likely that the priory was already committed to several corrodies that ate away at an institution's resources. Or the priory might not have entirely trusted Thomas Stafford's judgement in how he spent his money.

These 'jobs-for-life' and pension schemes outlined in the contract drawn up when the Master Mason was hired should not be confused with corrodies taken by other Master Masons when they retired, or were thought to be nearing the end of their working lives. A corrody was originally the benefactor of a monastery having the right to have lodgings in it, or to give that right to someone of his choosing. It came to mean other types of pensions and board and lodging given to someone who had given service to the monastery. There was also a system of securing a corrody by paying a lump sum and getting a kind of annuity for life in return from the monastery, the abbot and monks hoping (in the most Christian way) that the corrodian would not live too long. Sometimes this worked in the monastery's favour. Robert the Mason (he who was praised earlier in this chapter) was given a house and lands for his good service by St Alban's abbey church. Out of his own money, and perhaps from rents from the lands, he gave 10s a year to the abbey. On his deathbed in 1119, he gave back to the abbey the majority of what it had given to him. This speaks of a contented working relationship between monks and Master Mason.

A corrody might not be cheap. In 1316, William de Shockerwicke paid £60 in silver for his corrody for life from Worcester, which is an enormous sum considering that John le Roke, for example, was paid 6d a day in 1313 at Westminster (just over £9 a year, if he was paid every day, which would have been highly unlikely). For his money, Shockerwicke was given:

> a monk's loaf and a white loaf ... every day for life with two gallons of the
> best ale, a mess of meat every flesh day and two dishes of pottage such as the
> monks in the infirmary receive; supper equal to that of two monks; and on
> fish days what is served to a monk in the refectory.[25]

In addition, Shockerwicke had a room in the tailor's shop, a stable for his horse and whatever he needed for his servants. Whether or not he got value for money we will never know as his date of death has been lost. In 1364, John Sponlee was granted a corrody at Reading Abbey at King Edward III's request, having previously been awarded 1s a day for life for good service to the king. This is unlikely to have thrilled the abbot because Sponlee will immediately have become a financial and physical burden without having paid his subscription or, indeed, having given any particular service to that abbey. He lived until about 1382 so this was a substantial reward, very common among the Master Masons and indicative of the esteem in which they were held by their patrons. It was also a substantial drain on the abbey's resources.

Some of the payments outlined in the contracts may seem a little strange to us, but just as someone enlisting in the armed services today would expect to be issued with appropriate clothes, so giving robes of office to Master Masons was par for the course. There are records of this happening as early as 1171-2 when one mark (13s 4d) was allocated for Godwin the Mason's robes, and two years later Richard Fitzwiching had more expensive ones at the cost of £1. In the twelfth century, both sums would have been significant and as much as some men might hope to earn in a year, so this is yet another example of the high regard in which Master Masons were held and the expertise they had. Some men were given summer robes as well; others were given fur-trimmed robes for themselves and their wives, and what we do not know is how fiercely these details were thrashed out at the negotiating stage of the contracts. Were the wives who were given robes altogether more indomitable than those who were not? The Master Masons may well have been formidable characters on the building site, but whether they were as dominant at home, we do not know. Occasionally, the robe did not arrive on time and had to be sued for, which also shows that it was considered worth having by those to whom it had been granted. On one occasion that we know of, the robe was refused: in 1392, John Palterton flounced out, rejecting his robe because it was delivered late.

Other clauses in contracts included, in 1387, a 24ft (7m) measuring pole for Robert de Wodehirst at Ely, where he was also paid £4 a year plus his board and lodging. In 1516, Henry Redman was paid £5 a year at Westminster Abbey, plus a clothing allowance and 'glewe and sprig nale for Mr Redman to make his tools'. In 1415, John William and William Austeyn made the roof of Hartley Wintney priory, where the prioress agreed a sum of £22 for materials and labour, a pig, a wether, a gown worth 10s or 10s in cash, and said that she would feed John William and six men for a week when they set the roof in place. There is a slight air of exasperation about that contract on the part of the prioress. John Wyther and his men got so much food in 1267 that it is amazing that they managed to do any work. A staggering £6 was allocated for meat, bread, wheat, forty-eight carcasses of bacon, twenty-one cheeses and six casks of cider. It would be good to know how many men there were on site, or if there were only a few men accompanied with their extended families and bellies. Several men had contracts that granted an allowance of herrings, including Richard Russell, one of the few Master Masons who became an MP. He and his colleague, Adam Powle, were also to be paid by the yard as the steeple at Walberswick grew in height: one yard was worth £2 and a cade of herrings to them every year: a cade being a barrel holding 600 herrings – quite an incentive to work. The herring was a staple of the medieval diet, with larders in aristocratic houses having as many as 8,000 of them, together with six barrels of lard – the perfect diet! Richard Woodeman working, it must be said, at a slightly lower financial level than some of his peers, made a new floor for the belfry of Lydd church, in Kent, and was paid £2 6s 8d for all materials and labour and a good breakfast for him and his men.

The wealth of a Master Mason

As has been indicated earlier in this chapter, the potential for these men to become extremely wealthy was great, although it is extremely difficult to assess how much their wealth was worth in today's money. One suggestion has been that £1 in 1500 would have the buying power of £429 in 2000, but this seems too specific. It is quite hard enough for us in 2018 to recall what we could buy with a pound in 2001 without going back through changes of coinage over 500 years. Commodities change their value as well; gin and oysters were the food of the poor in Victorian London, but today would be considered expensive. A calculator was too expensive for most people to consider having at home in the 1960s, but now these are given away free in marketing promotions. In the twenty-first century, there are numerous tasks that, in effect, machines are employed to carry out. With luck, we will never have to pay an individual to empty a cesspit by hand, which cost 6s 8d some 600 years ago.[26] Nor do we have to pay someone to bring that most precious of commodities, water, in carts to our doors. If a man were given one shilling (five pence in today's money) in the first half of the fourteenth century, he might buy three gallons of best Gascon wine in London, or a sheep or half a pig, or half a brass pot, but at that time a carter or groom would only have earned 5–8s for the whole year.[27] In addition, the cost of living changed in the 500-year period on which this book focuses, which adds to the difficulty. Wages for the Master Masons were generally lower in the countryside than they were in cities, which was a normal pattern and applied to all trades. This meant that pay ranged hugely from men being paid a few shillings for a short-term job to, say, in 1174, when Maurice the Engineer was paid £18 5s a year, plus his robes. Other great architects, such as James of St George, were paid a colossal £54 15s in 1284. Hugh Herland earned £43 3s 4d in 1397, plus robes and a house, although the average panned out at anything between £18 and £36 a year. These wages also reflected what the patron was prepared to pay and it is certainly true that the richest organizations in the land were the Church and the monarchy; if they were working together as was often the case, especially in Henry III's time, so much the better. What is clear is that as soon as these men were able, they invested in property (frequently shops) and so increased their wealth through other businesses, rents and farming. This means that tables showing their wages often do not reflect the actual prosperity of an individual.

It should not be thought that money was always no object when it came to building for God, nor that every Master Mason made a fortune. There are several examples where the project ran so far over budget that it had to be stopped, such as at Nostell Priory in Yorkshire in 1328, where their ideas for a new choir were bigger than their purses. It was claimed that this was the fault of the designer, Robert de Pontefract, and possibly this was true. Richard Winchcombe refused to compromise on elaborate details at the Divinity School in Oxford and was promptly replaced by Thomas Elkin, who was more open to suggestion. These cases are rare, which says a lot either for the initial assessment of funds needed or for the tolerance of the patron – usually a bit of both. They would have allowed for some exceeding of the budget as we would today, though I doubt that they liked it any more than we do.

Over all of the 1,284 pre-Reformation Master Masons known to us, remarkably few seem to have got themselves into financial strife, and nor should they have done, given the wages they could command. Nonetheless, when as eminent a man as John of Gloucester, who was a friend of the king, died in 1260, he owed the king eighty marks, which was almost as much as the most successful earned in a year. He owned property, but only made two marks a year as profit. His problem was that too much had been thrown upon him and perhaps he had not learned to say no; the manager of the project needed to be very clear as to what he would or would not subsidize out of the wage he earned and he must have been too kind to his patron. It was difficult not to be when your patron was the king, but the incident with the loan of the wine casks suggests that Henry III was taking advantage of him. John Wastell's will in 1515 indicated that he had a similar problem because he had remarkably little to leave considering how he must have been one of the big earners of his day. It would have been by no means the first time such losses on a contract had been made: we know that, in 1450, John Goldyng suffered such great losses on a contract that he was paid compensation.

The two saddest stories are of Robert Carow and Robert Janyns senior. Carow died intestate in 1531 and it was found that his entire estate amounted to £2 7s 9d, which mainly went on his funeral and outstanding debts. Towards the end of his life his neighbours gave him food, some of which was paid for out of his estate

as well as 10d 'to a poor woman for keeping him in his sickness'; it sounds as if
it was a lonely end, even though his neighbours saw to it that he did not actually
starve to death. Janyns was a talented and accomplished craftsman who somehow
never managed to land that one big contract that would have been a springboard
to greater things. He ended up taking work where he could and, although he
had a regular appointment at Eton College, the pay was meagre. He found it so
difficult to make a living that his wife took up work as a carter, an unusual trade
for a woman in those days, but it brought 3d a day to the family on the days that
she could find work. His son and grandson did rather better but, in the end, the
Janyns family's situation became so bad that Robert senior has been obliged to go
into the Civil Service – a shocking end for a Master Mason.

CHAPTER FIVE

BUILDING CONTROLS

Maintaining the fabric of the building

GUIDE BOOKS OCCASIONALLY tell us the name of a few of the Master Masons responsible for a particular building, or the person who founded it, but that can only ever be one fragment of the picture. Norwich cathedral was commissioned by Bishop Henry de Losinga in 1096, but Richard Machun, Richard Uphalle, Richard Curteys, John Ramsey I, Thomas de Plumstead, William de Ramsey II, Hervey de Lyng, John Horne and William Reppys (to name just a few) are recorded as being actively involved in the construction and design of that same building over a period of 400 years. It is rare, if not impossible, to find a cathedral that has not been extensively changed, and even quite small parish churches show signs of significant alteration over the centuries. It is normal to be able to see where a window has been blocked, a porch added or the line of a roof altered. The glory of this is that we can frequently find styles changing in the same building, often in the form of a Romanesque crypt or nave, a Gothic choir, and perhaps Perpendicular windows or a chantry being inserted later. There was almost always work in progress, but it has been estimated that between 1350 and 1530 not only were some 1,000 new churches built, but that 2,300 either had towers added to them, or had them altered or heightened.

Structures of that age need constant care and maintenance from the outset, and this was catered for in the form of committees of inspectors called Expertises, which sought to ensure that the strongest possible structure was used in the first place. There is a difference, of course, between maintenance and growth, although they are often linked, since the collapse of one part of the building (such as the tower at Ely) might lead to the development of something new (in the case of Ely, the octagonal turret). This means that maintenance and growth were also closely linked to patronage. Thus, we have John Smyth, the rector of Knapton, in Norfolk, donating a fabulous angel hammer-beam roof in 1504. Also, Thomas Wikyng leaving not only a cow to the church of St Mary Magdalene, Cowden, but also 20s for '… the painting of Our Lady and repairing of her'. Finally, Avice de Crosseby of Lincoln bequeathing a very small leaden vessel to mend the eaves or gutter of the church of St Cuthbert in 1327; such a gift could be sold or melted down for direct use.[1] Patronage will be considered in more detail in Chapter 8.

In many ways, the maintaining of a cathedral was more straightforward than that of a parish church since the former was the continual headache of the Chapter who had masons and craftsmen on permanent strength or could find them fairly easily; the prestige of working on the cathedral was always helpful when it came to hiring labour. At parish level, the responsibility was split between the priest and the people, although it cannot be said that the congregation always felt very enthusiastic about this because they often had to be threatened with excommunication, if not imprisonment, to take up the challenge. The congregation had the upkeep of the nave, while the priest looked after everything east of the chancel step, which helped the Church with its financial outgoings. The Church probably thought it a perfectly reasonable charge on the congregation since the nave often doubled as a kind of village hall. Prior to the Reformation, it would have been unusual to have had fixed pews. It is rare to find evidence of formal seating for the congregation; St Mary's, Dunsfold in Surrey is believed to have the oldest set of pews in England dating to the thirteenth century, originally appearing to have been rows of plain benches without backs to them. In general, the nave would have offered an empty space that could also be used for indoor fairs or other community events. The people would have stood during Mass unless they were infirm, in which case they might have perched on low benches sometimes provided against the walls and still visible, for example, in the north aisle of St Peter's, Wolferton, in Norfolk. This is said to have been the origin of the expression 'the weak go to the wall'. There was a great deal of secular use of what we see as an entirely spiritual space, the church being naturally divided into two main sections by the rood screen that separated the nave from the chancel. It would have been very easy to make a more secure division had there been any anxiety that an indoor fair might get out of hand.

Nonetheless, the upkeep of the church was onerous and was seen as yet another burden on the people. Occasionally, if the parishioners were reasonably well off, they were able to set up a working arrangement, as they did at Glapwell, in Derbyshire, whereby they would give land that could be used to raise money as a kind of one-off payment for future work on the church. It should be said that it was originally thought that the parish should be responsible for the entire church, not just what is west of the chancel step; after all, it was a community asset. Legislation such as that passed at the Synod of Exeter of 1287 laid down who was responsible for what, and out of this evolved what we would recognize to be church wardens. Church wardens are still responsible for the maintenance of the church, among other things. The position arose from the need to have someone who could not only take charge of the maintenance and furnishings, but who could also be relied upon to coordinate the raising of funds. As today, this was an endless task and was mainly achieved by persuading volunteers to give additional money and/or to hold the medieval equivalent of the church fête known as 'church ales' (literally, ale being made and sold in aid of church funds – these days coffee mornings are more favoured). All of that said, while it was an unwelcome charge upon the parish, local churches were physically easier to maintain than cathedrals because mainly they lacked the soaring height and massive ground space of the latter, never mind the intricacy of carvings and glass and the press of visitors.

Building disasters

Medieval builders learned quickly from their mistakes; we are, after all, looking at their successes because their failures fell down. The dangers to any building, but especially to a very high one, were great: fire, flood, storms and poor building techniques were the main ones, and the Master Mason knew that while earthquakes and storms could not be avoided, he could anticipate the other events when he planned the building. Therefore, Master Masons often over-buttressed their buildings, although they did not know that at the time. Huge calamities still occurred, some of which came about slowly. We hear that at Ely cracks in the walls supporting the central Norman tower had been spotted and so the daily services were being held in another part of the building. This was just as well because when the tower collapsed on 22 February 1322, no one was injured, although the noise and destruction must have been quite frightening. The mess this made is best not dwelt upon, although when the tower fell at Gloucester in about 1170 right in the middle of Mass, it seems that 'so dense a cloud of dust from the shattered stones and mortar filled the whole church that for some time no one was able to see, or even to open their eyes.' Miraculously, no one was hurt because the tower was at the extreme west of the church, not over the central crossing, and the congregation had drawn nearer to the high altar at the east end for the bishop's blessing at the moment disaster struck.

As one of the factors in Ely's mishap, the monks had always claimed that they held the true bones of St Alban, which had attracted a lot of pilgrims and associated income. It was a genuine shock to the monks in 1314 when Edward II decreed that the true bones were actually the ones held in the abbey in the town of St Albans. Eight years later, Ely's tower collapsed: medieval man could see a connection that we may smile about until we remember that some people saw a link between Bishop David Jenkins' supposedly heretical remarks and the fire at York Minster in 1984. He had been consecrated bishop of Durham at York Minster on 6 July that year and the lightning bolt that did so much damage happened three days later (although why the disaster did not happen during the service itself has never been explained by the lunatic fringe). A similar incident happened at Lincoln in 1239, when it is said that destruction came about as a result of an argument between the canons and Bishop Grosseteste. One of the canons is supposed to have said in his sermon (possibly unwisely), 'Were we silent [about the bishop] the very stones would cry out for us' whereupon the tower crashed down, killing three monks and seriously injuring several others.[2]

St Alban's itself was not necessarily protected by its saint. In 1323 there was a terrible accident when two columns on the south side suddenly fell down. Unfortunately, a large congregation had gathered to hear Mass at the time, although it seems that people were more frightened than physically hurt. About an hour later, the roof that the pillars had supported came down, bringing beams, ties and rafters with it and crashing into much of the cloister below. The noise must have been frightful. Three days later, when the dust was still settling, there was an attempt to start clearing up the chaos when one man climbed up the heap of rubble to tackle it from the top. At that point, all of the remaining part of the standing wall fell, taking him with it. It seems that previously, at the end of the twelfth century, St Alban's had suffered the ministrations of one Hugh de Goldclif, a Master Mason described as being deceitful and unreliable, albeit a craftsman of great reputation. It appears that he let the costs mount up without significant results. The foundations were dug, but work did not progress. The walls were left uncovered in bad weather until the stone deteriorated so badly that carvings fell apart and columns fell down. Workmen were not paid and so left, leaving the poor abbot with a terrible mess on his hands. He resolved the situation by appointing Brother Gilbert de Eversolt as the overseer, allocating funds and generally keeping a much tighter grip on proceedings when a new Master Mason was found.[3] It must be said that the building of St Alban's has suffered numerous reverses over the years, but it still stands.

Beverley experienced two building disasters, the first taking place at the very end of the twelfth century. The account recorded at the time spoke of a very high tower of astonishing beauty and size that had stood for some 150 years. It still lacked a stone roof of proportionate height and this was added with more emphasis on art than on strength.

When they set up the four piers as supporters to carry the whole mass, they
let them into the old work ingeniously but not firmly, in the manner of those
who sew new cloth into old.[4]

In the end, gaping cracks appeared and some marble columns split the whole way
down their length so that people were rightly afraid to enter the building.

In October a priest, unable to sleep, rang the bell for matins an hour too soon.
When the canons assembled, a number of stones fell from the tower; they left
their stalls and, after a second fall of stones, went to the west end of the nave
and finished the service there.[5]

Moving was a good plan, but it was either brave or perhaps foolhardy of them
not to leave the building altogether. Not long after, the whole tower fell to the
ground. One account remarks that the craftsmen were:

not as prudent as they were cunning in their craft; they were concerned
rather with beauty than with strength, rather with effect than with the need
for safety.[6]

A second disaster occurred one Sunday in April 1520 when the central tower at
Beverley minster collapsed, causing numerous casualties.

At Beauvais, in northern France, the tower did likewise in April 1573, but
happily there a service had just ended, so the cathedral was empty. Towers also
went crashing earthwards at Worcester, Chichester and Winchester, some while
they were being constructed. At Winchester the cause was obvious to believers: William Rufus had been buried beneath it and he had been an evil king
– although it is not clear why a man who spent his eleven years on the throne
largely consolidating what his father, William the Conqueror, had begun should
be seen to be malevolent. From writs and decisions he made, we can see that
William Rufus was shrewd and sensible, very much a man of action. He was
pious and generous to the poor, but as a medieval king with all that that entailed,
he also made decisions that were unpopular with the Church. More boringly,
but no doubt more accurately, a contemporary chronicler put the Winchester
problem down to shoddy workmanship. So great was the danger of collapse that
a prayer was routinely said at the end of Evensong that went: 'deare Lord, support
our roof this night, that it may in no wise fall upon us and styfle us. Amen.'

There had been anxiety at Beauvais' cathedral, which still suffers from its
astounding height. The iron cross on the lantern had been removed in 1572,
the year before the collapse of the choir, in the slightly naive hope of lessening

the weight enough to prevent any disaster. This was not Beauvais' first building calamity since part of the vault had come thundering down in November 1284. Beauvais remains the highest of all the cathedrals, arguably winning the race for the sky. Inside, massive beams keep it propped up to this day, with extensive maintenance still going on. The vaulting inside the cathedral is over 150ft (46m) high, so it is all but impossible to see the remarkable stained glass at such a distance. Like it or loathe it, it is extremely impressive and much more so when it was first built; the citizens of Beauvais and anyone for miles around must have been astonished. The fall of even a part of such a leviathan must have shaken every house nearby and settled a thick blanket of dust all over the city.

Building neglect

At a lower yet just as catastrophic level, there is evidence in parish churches that maintenance often became too difficult, perhaps for reasons of money, will or manpower. The ruined transepts at Cley or the romantic-looking semi-ruin at Little Cressingham stand as testimony to that. Pershore Abbey in Herefordshire is described as a broken building, having suffered the effects of dissolution and structural collapse, and this could apply to numerous other of our most remarkable churches. Occasionally, human nature was to blame in the sense of work being held up disastrously because of argument. A sixteen-year dispute at a church in Leicestershire meant that there was such decay in the unfinished fabric that the chancel simply fell down in 1280. The task of overseeing the work had to be given to others to enable the job to be finished.

Beetles worked their way into wood everywhere; short-sighted birds have been a greater menace to acres of stained glass than any vandal; and, given the wind pressure some have to resist and the occasional abnormally violent storm, it is barely credible that any medieval glass survives at all. The Annals of Monmouth describe the terrible night the people of Winchcombe experienced on 15 October 1091; even given the medieval chronicler's delight in hyperbole, this was a very destructive storm, indeed:

> There were thunderstorms and whirlwinds ... a bolt from Heaven struck the tower of the church with such force that a hole was opened about the size of a man, entering by which it struck a great beam and scattered its fragments all over the church, and even threw down the head of the Crucifix with the right arm and the image of St Mary. At this time also winds, blowing from all quarters marvellous to relate, began on 17 October to blow so violently that they shattered more than 600 houses in London; churches were reduced to

heaps … The fury of the wind lifted up the roof of St Mary which is called at Bow, and crushed two men there. Rafters and beams were carried through the air, and of these rafters four of twenty-six feet in length when they fell in the public street were driven with such force into the ground that they scarcely stood out four feet and, as they could in no way be pulled out, orders were given to cut them off level with the ground.[7]

Guaranteeing standards in the building trade

We have already seen in Chapter 1 how expensive it was to build a cathedral and the practical difficulties of finding quarries and forests, as well as the preparation of the actual site to be built on. In the twenty-first century, we are accustomed to buildings being monitored throughout the construction process by buildings control officers, and this is by no means new. In the Middle Ages there was a system, known as an Expertise, that had the same function and which we see operating at every cathedral building site. Several full records survive of Expertises, which comprised groups of Master Masons not associated with that particular project who gave their expertise to assess how the building work was progressing. Sometimes they might be called in, as happened at Chartres in 1316, or it might be the result of a Visitation or formal regular inspection, as was the case at Troyes in 1362. Either way, there was always concern about the essential structure of the building. The Chapter would also be involved in this inspection, which was extremely thorough, albeit not necessarily welcomed by the Master Mason in charge. The following are extracts from the Expertise for Chartres. They begin by saying that some arches are good and strong and so no alteration would be needed, so there had evidently been some discussion about a particular part of the building. Then they list the problem areas, some of which go into minute detail:

First: we have seen the vault of the crossing; repairs are necessary there; and if they are not undertaken very shortly, there could be great danger.

Item: we have seen the flying buttresses which abut the vaults; they need pointing up, and if this is not done at once, much damage may ensue.

Item: repairs are needed on the porch piers and a plank should be provided in each side opening to carry what lies above; and, on the outside, one of the jambs will be moved above a reworking of the fabric of the church; and the plank will have a support so as to reduce the strain; and this will be done with all the ties that are needed.

Item: we have seen and devised for Master Berthaud how he will [re]make the
statue of the Magdalene where it now is, without moving it.

Item: we have looked at the great [south] tower and see that it has real need of
important repairs, for one of its sides is cracked and creviced and one of the
turrets is broken and coming apart.[8]

And so on. The Expertise of Troyes is similar both in its breadth and detail, and
these documents show how seriously these checks were taken and how much
Master Masons learned from each other. At Troyes, there is a record of the discussion about how much various repairs might cost, and an anxiety about the
maintenance passageways, which were filling with water. Cathedrals are riddled
with a series of passages within the walls at all levels; these allowed workmen to
access all parts of the building in shelter during construction, and still do. Some
of these inspection passages go right up through the clerestories (the very top
window levels of the buildings) and into the spaces between the vaults and the
roofs.

With modern technology, it is easy to forget or underestimate the achievements of the day. Girona's cathedral has a vault that is 73ft (22m) wide, which
was far greater than the norm when it was designed by Guillermo Boffiy in 1416.
The Works Committee of Girona did not think it viable and so called in no fewer
than eleven other Master Masons to give their opinion. Each man was asked
his opinion separately and each said that Boffiy's plan was both strong and safe,
which has proved to be true since the vault is still there. In these meetings, it was
not just the structure that was discussed, but the aesthetics of it all: having an
immense building that would resist gravity and time was one thing (and something to be desired), but it must also please both the eye and God. Records of
these discussions survive from Milan when it seems that some of the Master
Masons who were consulted were not convinced that if a building's proportions
were correct, then it must follow that it would be structurally sound. There were
other issues to do with the design at Milan but, essentially, those working on the
design around the year 1400 did not agree on what the basic proportions should
be, making it very difficult to proceed. It is worth mentioning that the cathedral
is still standing with all of its 135 spires.

In 1440 there was a real problem at the church of Rouen's Saint-Ouen, which
is cavernous and occasionally mistaken for the cathedral. The four piers of the
tower were carrying a vast weight, but were not supported by buttresses and so
they had started to bulge outwards. The fear was that even the slightest shift in
the building or any minor subsidence would cause the whole lot to topple. The
danger was perceived to be so great and money would be needed so rapidly for

repairs that they declared 'that it would be wise to sell or pawn chalices or objects of value to get money so that work could begin at once, and the church might be made safe'. Embarrassingly for the Master Mason in charge (Colin de Berneval), it had been his father, the late Alexander de Berneval, who appears to have been to blame for this costly miscalculation. Colin made sure that he could not be held personally responsible and asked for a copy of the report for his indemnity in the future, which is also interesting in that it shows that Master Masons could be (and were) held to account and knew they must take measures to safeguard themselves.

Towers and spires always created tensions in the fabric of the building and probably tensions among the builders and hierarchy, too. They were prerequisite for any church, but evidently placed so much stress on the structure that some-times they were built separately close by, such as the bell-towers at Chichester cathedral, Brookland in Kent, Beccles and Bramfield in Suffolk, Dereham and West Walton in Norfolk, Bosbury and Pembury in Herefordshire, Marston Moretaine in Bedfordshire, and many on the Continent. The tower and spire of Salisbury cathedral were added between 1297 and 1320, some half-century after the main cathedral building was finished. They make a sensational outline, but unfortunately weigh 6,500 tons (5,900 tonnes), which made the piers of the central crossing start to bend and sink. It seems unlikely that the original Master Mason imagined that anything quite so weighty would be added to his original design. Robert Wayte was called in to increase the strength in 1415; he did so by adding strainer arches. The tower and spire combined are 404ft (123m) high, making it the tallest in Britain.

Wells cathedral, in Somerset, offers a classic example of extensive maintenance being turned to good effect. The elegant scissor arches that separate the nave from the chancel were the inspiration of the Master Mason, William Joye (*see* plate 11). He had been appointed on 28 July 1329 at a fee of £1 10s 6d yearly for life, which was later increased to £2 yearly, with an extra 6d a day for each day that he was actually at work on the cathedral. His remit was to 'oversee repairs in the said church and give his counsel, care and aid thereupon'. Again, as at Salisbury, the problems began when, in 1313, an original central tower, which had been built over a hundred years earlier, was heightened. This caused the foundations of the western piers to subside about 4in (10cm), which in turn cracked the tower down the centre. The scissor arches that Joye inserted act as a brace, and they are not only lovely to look at but also effective, since there does not appear to have been any further movement since he finished the work in 1348.

Towers and spires inevitably do shift a little as age settles them and they are exposed violently to the elements. That said, Norwich cathedral's Norman tower has only decreased in height by an estimated 9in (23cm) over one thousand years.

The spire it supports is not only the second highest in Britain, with a combined height of 313ft (95m), but is also the fourth spire in that position because its predecessors succumbed to fire, lightning and storms. The spire we see dates to about 1480 and is the subject of continual maintenance to this day.

Medieval building inspectors

Along similar lines to the Expertises were the groups of men known as Viewers and Searchers, who crop up in the records of Master Masons. We might recognize in them the role of district surveyors and their history appears to go back to before 1300 when qualified men were enlisted to look into a problem and to report back to the mayor or aldermen of a city, thus aligning the building function to the civic function just as it is today. No town or city can operate successfully without rules followed by all parties. The Viewers and Searchers were appointed by the city, not by the various trade hierarchies. They were also known as Sworn Masons or Sworn Carpenters since they had to swear to give:

> proper consideration to all men of the city and suburbs … concerning ruinous, partible and non-partible stone walls between neighbours and touching other things pertaining to his office, as often as required to do so.[9]

The term Sworn Mason could also indicate that someone had been sworn into a particular guild or lodge. Searchers existed in every craft and, indeed, William Vesey, a Master Mason, was granted the office of Searcher of beer-brewing in 1441 for good services to the king, which suggests that there was a strong financial benefit attached to the appointment. It is assumed that brewing was not his forte since he was a mason by trade. The Viewers' role was to keep up standards and to search out infractions. They were not only required to check building sites, ensure that building and labour codes were being followed and report any bad workmanship and/or bad materials, but they also had to arbitrate in arguments over boundaries or any other planning type of considerations. In 1457, the Master Mason, John Porter, was called in with his three colleagues to adjudicate in a dispute that had arisen because poor drainage meant that rainwater from the yard of one house was spilling into that of another. This example suggests that some of these arguments were perhaps more civic than architectural, but they had to be resolved all the same. The officials might also act as intermediaries between the workforce and the city since they were responsible for maintaining effective labour relations. In essence, two masons and two carpenters at the top of their craft in most cities held the office of Viewer or Searcher at any one time, usually for the period of a year, and they worked together throughout as opposed to being called in for the specific, perhaps one-off inspection of the Expertise. Numerous Master Masons fulfilled this role, to the extent that we can see them working together in cities such as Norwich, York and especially in Bristol; sometimes, this even helps to verify dates of other events in their lives. The carpenter Richard Somere and masons John Walter and Ralph Sporrier appear to have worked together in this capacity for the city of Bristol between 1493–6, but the appointment of Viewer can be found nearly 300 years earlier. Jordan the Carpenter and Engineer made the king's trebuchets at Dover and later at Windsor. Jordan was appointed King's Carpenter and even King's Catapult-maker (which was worth 12d a day) and was made Viewer of the works at Windsor Castle in 1227. Most Viewers led more conventional careers, but the fact that so many are recorded in this role, and often more than once, indicates that this was at least a useful career move for the individual concerned. It must have been lucrative or it would have been hard to get anyone to do it, such is human nature. It also shows an early and ongoing interest in what we would call health and safety.

These appointments are not likely to have endeared them to their fellow masons and carpenters because reporting on the poor workmanship of their colleagues was a major part of the job. Despite the marvellous standard of building that we see today, the fact that there were catastrophes does show that not every building's structure was thought through carefully, as has been seen at Rouen. Meaux cathedral was ripped down altogether in 1197 because it was 'unsatisfactory in plan and construction', and the new east end of Durham had 'as many fresh starts as masters' around 1153.[10] The ultimate design of a cathedral was not set in stone, so to speak, from the outset so it could easily result in an end-product that might need a stronger basic structure. This was partly what the Expertises, Viewers and Searchers hoped to eliminate, but inevitably some workmanship was so shoddy that it ended in a lawsuit, more usually concerning a house than a church. In 1317, John of Bytham was charged in the Fair Court of St Ives for having put alder and willow into a house when he had been contracted to use oak. He pleaded not guilty. We do not know the verdict, but, as it would have been easy enough to discover the truth, it seems unlikely that the case would have gone to court unless he was guilty. At the end of the fifteenth century, John Tilley and Richard Cuttyng were sued for defective work at the church of Little Thornham, in Suffolk. William de Hoo, with three other masons, was contracted to make a wall around the manor of Eltham by Michaelmas 1316 to the design of, and under the direction of, the Master Mason Michael of Canterbury. Perhaps Michael was an absentee supervisor, for the workmanship of de Hoo and his colleagues was held to be fraudulent on account of them failing to build the wall to the specifications laid down. They also used inferior materials with the result that they were imprisoned and heavy damages awarded against them. They were released on an undertaking to rebuild the work properly with Michael of Canterbury, Alexander Le Ymagour and other citizens paying surety for them, although it is not recorded how much persuasion they had to use on Michael to do this since he must have been annoyed, to say the least, to find that his early trust had been misplaced. This will have damaged Michael's reputation as much as theirs and one can imagine him supervising more closely the second time around.

John Mason of Seighford, in Staffordshire, and Nicholas Mason of Stafford were sued for damages of twenty marks (£13 6s 8d) by John Harecourt in 1479 because they had been contracted to build a tower at Swynnerton, but had constructed it so badly that it had fallen down. From the wording of the case, it sounds as though the men had been paid in advance – never a wise move with builders, in any case! In 1473, Robert Sherbourne undertook to do some work in the chancel of High Wycombe church, but twenty years later he still had not done so, and it was deemed that the chancel was ready to collapse. It is not recorded if this was because of idleness on his part or because he had taken on too many other jobs, as often happened with other Master Masons. Those cases apart, there are remarkably few surviving complaints against the Master Masons, re-emphasizing the level of excellence they had to achieve to get to the top of their trade. All that the lawsuits prove is that there were exemptions to every rule and every trade was blessed, or otherwise, with people who had somehow won through despite being either deceitful or not up to the job, and it will ever be thus.

The Viewers were neither working in the dark nor making up rules as they went along, nor indeed merely hoping to spot an error that they personally would not have made. They worked from experience and common sense, but they were also checking materials and work against a set of standards, some of which survive to this day, although mainly the extant documents concern the regulation of guilds and commercial competition. Badly prepared stone could become waterlogged and subsequently crack in winter frosts, so building sites would generally close down for the winter, the walls having been thatched to help protect them from the elements.

The working day on site

During the ten or so months of a year when building could take place, work might stop because of various festivals and also in accordance with religious services. In Paris, work was supposed to stop on Saturdays when the bells rang for the religious office of None at any time of the year (mid-afternoon), or after Vespers at about six in the evening unless the builder was about to lay the last stones of a doorway leading into a street, or if he was about to close an arch or a stairwell. Anyone defying this rule and working longer hours could be fined four deniers by the guild master, who could even remove tools from those who regularly transgressed which, of course, meant that they had lost their means of earning a living. There were roughly defined hours of work, which were considerably longer than modern ones. That the working day started comparatively early is shown by a regulation from Coventry of 1553, which says that:

> All Carpenters, Masons, tilers, dawbers and also all kind of labourers within this City lacking work shall assemble themselves at five of the clock in the morning in the summer time with their tools in their hands at the Broide-yate according as in times past they have done to thentente such as lack workmen may find them there, and that none of them be found idle at home or in any ale house.[11]

And this from the Rules for the Conduct of the Masons at York Minster of 1370 is even more specific:

> It is ordained by the chapter of the kirk of St Peter or York that all the masons that shall work on the works of the same kirk of St Peter shall from Michaelmas Day unto the first Sunday of Lent be each day in the morning at their work, in the lodge that is ordained to the masons to work in with the close beside the aforesaid kirk, as early as they may see skilfully by daylight for to work. And they shall stand there truly working at their work all the day after, as long as they may see skilfully to work, if it be all workday; otherwise till it be high noon smitten by the clock when holy day falls at noon, so that it be within the foresaid time between Michaelmas and Lent. And in all other time of the year they may dine before noon, if they will, and also eat at noon when they wish, so that they shall not stray from their works in the aforesaid lodge at any time of the year at dinner time, but such a short time that no skilful man shall find fault in their absence. And in time of meat, at noon, they shall, at no time of the year, stray from the lodge nor from the work aforesaid, more than the space of the time of an hour, and after noon they may drink in the lodge. And for their drinking time between Michaelmas and Lent they shall not cease nor leave their work passing the time of half a mileway.[12] And from the first Sunday of Lent until Michaelmas they shall be in the aforesaid lodge at their work at the sun rising, and stand there truly and busily working upon the aforesaid work of the kirk all the day until it be no more space than the time

of a mileway before the sun set, if it be a workday; otherwise until the time of noon, as it is said before, save that they shall between the first Sunday of Lent and Michaelmas dine and eat, as is before said, and sleep and drink after noon in the aforesaid lodge. And they shall not cease nor leave their work in sleeping time passing the time of a mileway, nor in drinking time after noon passing the time of a mileway. And they shall not sleep after noon any time but between St Helen's Mass and Lammas.[13] And if any man stray from the lodge and from the work aforesaid or make any default any time of the year against this aforesaid ordinance, he shall be chastised with the reduction of his payment, at the supervision and discretion of the Master Mason. And all their times and hours shall be ruled by a bell ordained for this purpose. And also it is ordained that no mason shall be received at work, to the work of the aforesaid kirk, unless he be first proved a week or more upon his well working, and after that he is found sufficient of his work be received by the common consent of the master and the keepers of the work, and of the Master Mason, and swear upon the book that he shall truly and busily at his power, without any manner of guile, feints, or other deceits, hold and keep wholly all the points of this aforesaid ordinance, in all things that concern him, or may concern him, from the time that he shall be received at the aforesaid work as long as he shall remain a mason hired at work at the aforesaid work of the kirk of St Peter, and not go away from the aforesaid work unless the masters give him leave to depart from the aforesaid work. And whosoever shall come against this ordinance and break it against the will of the aforesaid chapter, may he have the malediction of God and St Peter.

In fact, most people operated from first light until dusk, so five in the morning would not have been so remarkable then and the hours would have been less than those worked by the average labourer involved in agriculture. Some early regulations indicate that masons working in England were allowed a siesta, albeit a short one and only applicable at certain times of the year. They had to prove themselves to be good workers before they were given a proper contract and there is an emphasis on preventing them from wandering off or slacking. It is also plain that they worked very long hours.

The dangers of fire

Having finished their daily work, they might then be required to take their turn at community duties, such as fire-watching. There are several surviving bridges that have high towers on them, such as at Monmouth, Cahors and Orthez, which

◁ Abbaye aux Dames, Caen – an example of the Romanesque style

Salisbury ▷ cathedral – an example of the Gothic style

◁ *Buttressing as seen at Amiens cathedral, Picardy*

△ *Tracery on a damaged window at Soissons, Hauts-de-France*

The hard-working oxen ▷ at Laon

◁ *The tomb of William de Wermington at Lincoln cathedral*

The tomb of Hughes Libergrese at ▷
Rheims

◁ *The tomb of Adam Lock at Wells cathedral*

The likeness of Henry Yevele ▷
on this boss at Canterbury was
probably taken from his death
mask

◁ *The font at Bridekirk*

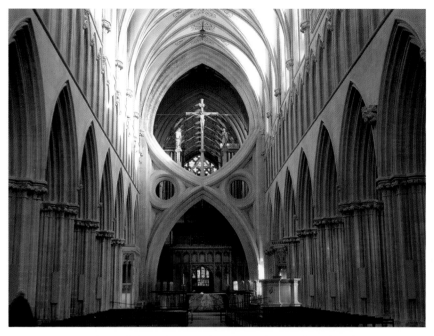

△ *The beautiful and functional scissors arch at Wells cathedral*

Rheims cathedral ▷
– an inspiration
for Henry III's
remodelling of
Westminster Abbey

△ *Westminster Abbey*

△ *The font at Stradbroke*

▽ *Arlingham church*

◁ *Putlogs*

Detail from the ▷
façade of Nimes
cathedral, showing
masons at work

◁ *Detail from capitals in St Mark's*
Square, Venice, showing sculptors
at work

were not primarily to control access to the town (although that too, no doubt), but for spotting uncontrolled fires, which were a constant threat. Fire was the greatest menace to any medieval building, as we have already seen at Chartres, even though the cathedrals were built in stone. Nonetheless, inside the building, candles were much in use and the walls were often covered in tapestries and other hangings that would have been flammable. Fire-watchers were put on duty especially if there had been a long dry spell and curfews were imposed in the original sense of the word (a merging of the French *couvrir* 'to cover' and *feu* 'fire'). Some fire precautions survive from 1423 Norwich, which detail how many fire-watchers there were to be in each ward and that their duties would run from the curfew until three in the morning, by which time, it is assumed, all fires had gone out and there was no likely risk from a late flare-up that might cause a rogue spark.[14] Only one person per ward could 'proclaim with a kindly voice' that everyone could well and securely keep fire and light in their places. Anyone refusing to take their turn for this duty was to be fined 6d and, if it should happen that anyone, day or night, should 'make affray by fire', he was to pay 100s or be imprisoned for a year.

So many towns and cities suffered from fires that we find churches now that seem to be sited well away from the parish they serve and it is sometimes speculated that this was because of plague. It may well have been that the people fled from the disease and then did not want to return to those buildings and so built new ones, but it was more common for fire to have destroyed the neighbourhood, which was then rebuilt further away. Some towns such as King's Lynn ruled that newly built houses had to have tiled roofs rather than thatch to try to prevent fire from spreading. Therefore, bridge towers were originally for the townsmen to take turns looking out for signs of unaccounted-for smoke, as well as to make a civic statement. In Dinan, in Brittany, the bell-tower doubled as a fire-watching platform, the river being too far away in a valley for the bridge to be practicable.

Men of all trades were posted on guard duty at nights, not just in case of fire, but to help to reduce crime. There were exemptions that, in 1258, included mortarers who 'have been exempt from guard duty since the time of Charles Martel [who died in 741 and was the grandfather of Charlemagne], as wise men have heard it said from father to son.'[15] There is no clarification as to why, although if the mortarer got his job wrong the whole building (and, therefore, lives) were in peril, but it seems unlikely that the early Carolingians were so interested in health and safety that his fatigue would have been of concern. Perhaps that is unfair because there was concern about safety on the building site: no one wanted to be hit by falling masonry or tools; nor did they want anyone's life to be ruined or squandered through carelessness (or any other reason). Even the lowest worker on site had value and was an asset. We can see some of this concern in directives of the time, although some appear to have been more interested in making sure that the job was done within a set time and to a set standard, as stated in this extract from the London Regulations for the Trade of Masons of 1356:

> In the first place, that every man of the trade may work at any work touching the trade, if he be perfectly skilled and knowing in the same ... Also, that no one shall take work in gross, if he be not of ability in a proper manner to complete such work; and he who wishes to undertake such work in gross, shall come to the good man of whom he has taken such work to do and complete, and shall bring with him six or four ancient men of his trade, sworn thereunto, if they are prepared to testify unto the good man of whom he has taken such work to do, that he is skilful and of ability to perform such work, and that if he shall fail to complete such work in due manner, or not be of ability to do the same, they themselves, who so testify that he is skilful and of ability to finish the work, are bound to complete the same work well and properly at their own charges ...[16]

Examining these clauses, it does seem as though they arose as a way of settling previous arguments and it is also clear that they could apply to any trade, not just the building ones. Later on, Jost Dotzinger, the Master of the Works of Strasbourg cathedral, convened a conference of master stone-cutters in Ratisbon in April 1459. His aim was to make the regulations of their various lodges uniform so that they were all working to the same standard.[17] What became known as the Statutes of Ratisbon were comprehensive yet practical, and will be looked at in more detail in the next chapter. Regulation Twelve states that 'the masters should do their work in such a way that the building built by them be without flaws for the length of time that is determined by the customs of their region'. Unfortunately, there is no explanation attached, but perhaps there was discussion about how to fend off the chance of complaints being made years later and whether or not they might be justified.

The guilds

There were some regions across Europe that must have been known for their poor bedrock, requiring extra-deep foundations, or for being especially prone to damp or strong winds. Some of the regulations may sound obvious, such as the ones that require only competent persons to be hired and, of course, that they must be competent themselves. The fact that this is included indicates that there had been a problem, which is interesting in itself considering how keenly the Master

Masons had to compete to get jobs in the first place, and how much depended on personal renown. If a Master Mason was brought in to finish someone else's job and turned out not to be capable of doing it, then there was to be an inquiry to ascertain what had gone wrong, followed by punishment, if appropriate.

As time went on, guilds became prominent in monitoring trades associated with building sites. Although the rise of the guilds as regulating bodies tended to be a post-plague phenomenon, they did exist before then. The early craft guilds go back as far as 1129 and included those for fullers at Winchester and for weavers in London, Lincoln, Oxford, Winchester, Nottingham and Huntingdon. These are static professions and so it was easier to establish clubs, where fullers and weavers might gather to socialize and discuss their trade. The original function of guilds, going back even further, was very different. The word comes from the Old English *geld*, meaning 'money' or 'payment', and in the early days implied a group of people who were not necessarily family, but who were jointly responsible for fines imposed on a community. These date back to at least the reign of King Ine of Wessex in the seventh century, but are not related to the later guilds any more than the Masonic movement now is a direct descendant of the stone-cutters' and masons' lodges of the Middle Ages. That said, the modern-day Masons share some of the secrecy of the medieval ones, who were trying to make sure that no one stole their ideas. Stone-cutters' lodges, built on a medieval building site to offer coordination, shelter and training, were well known for only allowing their own kind through their doors to preserve their professional dominance in the field: if you were not admitted, you were not part of the team and, therefore, you were neither trained or paid. It was an elite professional club that controlled access to employment and disciplined its members, mainly in the form of imposing fines. There is a particularly nicely worded clause in the Regulations concerning the Arts and Crafts of Paris of 1258, which says of quarrelling guild members:

> The guild master may only levy one fine per quarrel [so he decided who was at fault]. If the one who has to pay the fine is so furious and beside himself that he refuses to obey the order of the master or to pay his fine, the master may bar him from the craft.[18]

This would be an extremely severe punishment and on a par with the case of Richard de Northbury who lost his free status (see Chapter 4), as it would, in effect, put the individual out of legitimate business. These regulations also go into detail about how workers must not swindle each other perhaps by delivering underweight loads to a building site, or by cheating on the quality of materials. For instance:

If a plasterer mixes material with his plaster which should not be used, he must pay a fine of five sous to the guild master every time he repeats the offence. If the plasterer makes a habit of cheating and does not improve or repent, the guild master may suspend him from the craft; and if the plasterer will not obey the guild master, the latter must inform the provost of Paris, who in turn must force the plasterer to abandon the craft … Mortarers must promise under oath and in the presence of the guild master and other masters of the craft that they will not make mortar of other than good binding material and, should they make it of other material so that the mortar does not set while the stones are being placed, the structure must be undone and a fine of four deniers paid to the master of the guild.[19]

One can see from this how basic safety precautions were put into place, since good mortar is crucial to the longevity of the building and to the people building it as already indicated. It may be surprising to note that a mortarer's apprenticeship lasted for six years. We can also see how powerful a person the guild master was. The guild master was not part of the civic hierarchy of the town, being separate from the mayor and aldermen, even though they worked in support of each other.

Guilds of merchants came to the fore early on as a result of townsmen trying to protect their commercial interests from local town or aristocratic interference. Guilds of crafts quickly followed and, as has been seen, their role was several-fold: standardizing training and ensuring high-quality workmanship, establishing minimum prices and wages, regulating hours of work, preventing poaching not only of apprentices but also of projects, offering mutual support and restricting competition. Inevitably, workers within individual guilds also began to congregate together so that there might have been whole streets of silversmiths or weavers, which can still be seen in road names, such as those in places like York, to this day. That did not apply to the building trades because, as mentioned before, they tended to be a nomadic group, building a temporary lodge and moving on when the project was finished. That does not detract from the importance all guilds had in the development of town life, especially after plague reduced the available manpower for all trades and crafts and increased the negotiating power of the people. Guilds became an integral part of town administration. They were also a significant factor in the development of churches and cathedrals, partly because of their wealth and proclivity to donate anything from stained glass to chapels, but also because of the active role they played in the pious practice of the people. They organized miracle and mystery plays, sponsored religious festivals, ran charitable works and paid for numerous clergy. Some guilds employed priests to pray for them, buying land to fund them and, in this way, became as much a part of the local church or cathedral as they were of the town. Guild activity in the sense of patronage helps us to see how involved even ordinary people were in the Church and vice versa. Indeed, any form of patronage shows how important it was not just as an institution, but as part of everyday life and beyond.

Medieval attempts at monitoring and building control might raise a few eyebrows today, but there is no doubt that it was held to be important even if it was sometimes addressed rather late in the construction process. Very often it came out of disaster on a scale that we could hardly bear to imagine when we see the finished product. It may not have been as coordinated as we might like but between the Expertises, the Viewers and Searchers, individuals such as Jost Dotzinger, not to mention the leaders of guilds, there was certainly a distinct desire to make sure that the buildings were strong and ready for long service and that those working on and in them were safe. They may have been working to the glory of God but, being mere mortals, no one was anxious to meet their Maker ahead of time.

CHAPTER SIX

TRAINING AND BEYOND

Training the next generation of Master Masons

As is always the case, those who were at the top of the tree were continually on the lookout for youngsters with the talent, ability and perseverance to follow them; young men (usually teenagers – often as young as fourteen) who committed to train under one master until they themselves were ready to take their skills on the road to start working on some of our greatest buildings. It worked both ways: the youngster needed the training and subsequent career; the Master Mason needed someone to labour while gradually assisting with more complex tasks. We know something of how this worked through a handful of surviving documents – the *Regius Manuscript*, the Statutes of Ratisbon (1498) and the Statutes of St Michael (1563) – some from Germany, which have largely formed the basis of this chapter.

Training an apprentice was not something to be undertaken lightly by a Master Mason because both parties would be committing themselves to a number of years. The London Regulations for Masons of 1356 say seven years, but the 1563 Statutes of St Michael say that 'No craftsman may take on an apprentice for less than five years'.[1] Admittedly, the actual craft is not specified, but these are regulations for masons' lodges in general. Obviously, individual crafts, and not just those within the building trade, needed different levels of skill (the training of saddlers' apprentices lasted ten years). Only a few documents survive that might constitute a manual, so we do not know the precise details of any apprentice training and the system of indentured apprenticeship in any trade was not recorded until the late thirteenth century. For most, training will mainly have been carried out working alongside the expert, as from time immemorial. No doubt apprentices started with menial duties such as cleaning and worked their way up as individual masters saw fit, probably in accordance with whatever form their own training had taken. Knowledge was, therefore, likely to have been empirical rather than theoretical, although we do know from numerous sources, including Mathes Roriczer (of whom more in Chapter 9) that they were taught geometry, the 'craft of Euclid'. In broad terms, at the end of his period of training, the aspirant would have had to have proved himself capable of working unsupervised in all aspects of the trade and to have produced a proof-piece (the word 'masterpiece' does not appear until the sixteenth century). It is not clear how this was managed for masons, stone-cutters and the like, although they must surely have been subjected to what we would call continuous assessment.

In Paris the apprentice had to study for a minimum of six years and had to be paid a wage if he then stayed on to work with the same master. He had to do the full length of his apprenticeship and might only be released from his training earlier if he were the legitimate son of the master; perhaps that assumes that he would have learned more at his father's knee as he grew up. If that rule was broken and he was not paid, then the master would be fined by the guild, so it would have been easier just to give the lad his dues in the first place. Master Masons were also limited to having one apprentice at a time and supposedly were not allowed to take on a second until the first had completed five years of his training. That rule seems to have varied locally (to the extent of being ignored) and some were allowed an apprentice on each building site that was being overseen by the Master Mason so long as there were not more than five of them. The fact that Master Masons were limited in the number of apprentices they could take on also suggests that the learner was subject to close supervision and on-the-job training, which would not have worked if he had been part of a large group. This was one-to-one teaching with the master passing on his many secrets or tricks of the trade to a trusted novice.

That some of the Master Masons were also freelance teachers is revealed by a regulation that says that they were not to do this for money: if they wanted to instruct likely lads who were not apprentices and, therefore, were not in the system, then the Master Masons could do so, but only without charge. It seems that many tried to get around these rules by having numerous apprentices mas-querading as servants and general helpers, which led to masters being forbidden to instruct these helpers in any of the more detailed aspects of their trade. This again indicates an anxiety that the instruction could be diluted in some way that might have a disastrous effect later on the structure of buildings and, dare one suggest, even more so on the reputation of that particular lodge. The reason for a master wanting more than one apprentice was because he was paid a fee to carry out the training, but the arrangement would have had to have been above board for him to receive the money, so the extra boys were mainly acting as useful

cheap labour. There is varying information as to whether the apprentice was
paid a wage of any sort and/or whether he had to pay a fee before he was taken
on by the master: probably both is true. Certainly, the great Master Mason John
Wastell's son, Thomas, was recorded as being paid 1s a week as an apprentice
when his father died in 1515.

On the one hand there is evidence that apprentices bought into their training:

> A person who accepts an apprentice must not do so without charging a security
> payment of any less than twenty florins, which he must deposit with someone
> who lives in the same place, so that if the master should die before the end
> of the apprenticeship, the apprentice may serve the corporation with another
> true master and may complete the five years. But if he does not complete
> this period, he will give up the twenty florins to the corporation to cover his
> expenses and losses.[1]

On the other hand, the apprentice was to be paid both in cash and keep so,
having secured his place at his master's side with a down-payment, some of the
money was trickled back to him for his daily needs.

It seems that the master might not always be free to choose his own apprentices either, possibly to avoid nepotism since any vacancies for training would
have been much sought after. On 1 October 1471, Thomas Peyntour was granted
a patent for life as Master Mason to the cathedral priory of Ely. He was to receive
eleven marks a year (£7 6s 8d), a suit of clothes as provided for a gentleman
of the prior's household, and a fortnight's holiday. He was also to train three
apprentices, one after another, for the works of the priory. If these apprentices
were to follow the official timeframe in starting their training, this implies that
Ely Chapter expected Peyntour to be working for them for ten years and ideally
for seventeen to see all three youngsters through, which was not only a valuable
contract, but also a personal accolade. John Bell's contract was even more specific
in April 1488 when he was appointed 'special mason' to the prior and Chapter
at Durham for life: he was not to depart from his occupation without special
licence, he should have one apprentice of his own, and he must train another to
be appointed by the prior. This second one was probably a child either taken in as
an orphan or given to the priory as an oblate, a child dedicated to the monastery
by his parents and placed there to be brought up. (It may sound harsh, but it gave
the child the opportunity of an education and, therefore, a chance in life, even if
his childhood came with its own set of restrictions.)

Some evidence comes from other trades; for example, a potential cordwainer
had to be admitted to his training in front of the mayor and be shown to be of
good character, stipulations that applied across all trades. He had to pay 2s to the

City and another two to the poor box of the Cordwainers' Craft. In addition, he had to pay either 20s or 40s as a premium, depending on what sort of cordwainer he was to be, so only the sons of the well-off could apply. It does appear that once he had completed his training, any apprentice could find himself in financial debt to his master, not dissimilar to students leaving university now, and they were not supposed to take work with another master until that debt was repaid.

Regulating the training process

Likewise, an apprentice could not transfer to another master and a master could not poach from someone else. We can see how seriously this was taken by an entry in some London muniments, dated 21 July 1409, concerning John Dobson, who was a Master Carpenter at St Paul's cathedral. On that day, the sheriffs of London were ordered to release John Massingham who had been imprisoned for quitting the service of John Dobson before the expiry of his term of apprenticeship. Dobson must have died at this time so Massingham had had to seek patronage where he could, which must have been an all too familiar problem in those disease-ridden days. This Massingham is likely to have been the same man who went on to become a very successful carver and Master Mason in his own right, making statues for Canterbury cathedral's pulpitum screen and becoming the Master Carver at All Souls College, Oxford, among other high-status appointments. He was also the man who first introduced the use of landscapes in the backgrounds of sculpted groups in England, which had been pioneered by Claus Sluter on the Continent. This was a significant change in artistic terms at the time and a break with traditional ways of displaying biblical scenes.

A master had to be a freeman to take on an apprentice in the first place. It would be strange if a Master Mason was not free and, indeed, following in his footsteps was a route to freedom for the apprentice. Nonetheless, several cases of abuse were recorded whereby the so-called master failed to register his student at the outset of his training, and then did not present the successful candidate to the relevant authorities to be given the freedom of the city in his own right, thereby hindering his future employment and prospects, and defrauding him. It is not clear how the master got away with this because the guilds existed to regulate these matters. That said, it is likely that a mason or carpenter could get around these parameters more easily than other trades because they were continually on the move, the building trades being the most itinerant. According to the *Regius Manuscript* written in England in 1390, a person could not become an apprentice if he was deformed or lame since it would do the craft no good if the man was not capable of doing the work; he should also be of 'lawful blood'. He could be replaced by another candidate if he turned out to be too dim-witted or lacking in ability to become skilled at the craft. This was an issue that raised its head several times in regulations simply because the profession needed physical strength and intelligence. The eighth article of the *Regius Manuscript* also stipulates that:

> When a mason has not been able to go from being an ignorant to a skilled man, the master may replace him with someone who is better at his work. The clumsiness of the weaker one might do great damage to the image of his profession [and 'damage to the strength of the building' might also have been added].[2]

This must have caused a lot of stress between master and apprentice and it seems strange that clumsiness was not spotted before the apprenticeship even began, since lads were not taken on a whim. It must have been even more awkward if a Master Mason had taken on a trainee as a favour to a friend or family and then found him not to be up to the work. Curiously, although guilds laid down regulations for training and trading standards, they do not seem to have given written instructions in the sense of producing a syllabus, or even a list of basic objectives to be reached. It is assumed that he would learn how to draw a basic design and how to build not just walls, but archways, stairs and the more intricate details needed for tracery. We assume this, but we do not specifically know: the practicalities of training appear to have been left very much to the individual Master Masons. It may be that this meant that it was a gamble for the would-be apprentice who was about to commit to years of training when he was too young to understand the implications. Could he trust his new master to train him well or not?

What does come down to us is a clear understanding of the type of person who could be an apprentice and, later, a Master Mason. We can also see how the lodge, which began life as a safe place to leave tools, a place to discuss progress and to take a break, and a place to provide some sort of shelter, became an institution in its own right. Much of this progression be gleaned from regulations concerning apprentices, although many of the surviving documents were not written down until the latter part of the Middle Ages. The Regulations of Ratisbon, drawn up in 1459, make some basic provisions such as that no one can receive training of any sort if they are not part of the corporation. This does give an impression of a closed shop but, more importantly, of a central controlling body. The regulations also state that apprentices must be trained by a Master Mason, which may sound obvious but later we find that if a journeyman (or 'companion', as they were often known in France) went to the workshop committee and asked to be hired, his request could not be granted unless the person with whom he had served his apprenticeship was himself a Master Mason. This meant that no one could be trained casually with the hope of getting prestigious employment through a back-door later on so, even if the Master Mason had been freelancing, it would not have done the student any good in the long term.

One of the early regulations states that the apprentice must be married, even though the individual must often have been well below the age where he might have a wife. This rule is contradicted by a later one that allows an apprentice to be released from his indenture if he were to get married. The reason for this is not revealed, but one assumes that the marriage meant that the trainee had to move to a different area, which also implies that he had chosen his bride well and that it was better to be nearer to her family than his own. There is reason to believe that that was the case because the great Sicilian Renaissance artist, Antonello da Messina, had an apprentice called Paolo di Ciacio from mainland Italy who reneged on his contract after a year on the grounds that he had got married and wanted to be with his bride. Antonello took him to court where it was agreed that Paolo would return to his service until he had paid back the money and goods provided for him during that first year of training.

More surprisingly to modern readers, the Ratisbon regulations say that it must be ascertained that the potential apprentice was not illegitimate: 'No master or contractor may hire an apprentice who is not married. In addition, it is right to ask him whether his mother and father are married',[3] which might be regarded as an impertinent question today. The Ratisbon document goes on to say that if an apprentice behaved badly in his personal life and especially in his relations with a woman outside marriage, he should lose the profits of his years of apprenticeship. His case must nevertheless be examined with understanding – that last statement revealing that the hierarchy of the guild may have remembered their own debauched youth. The regulations indicate that the apprentice should be someone of a good family anyway or, at any rate, not a serf. This may sound harsh to modern ears, but it was to do with education (or the aptitude for education), money and a presumed innate understanding of how to behave, not necessarily morally, but in terms of business. A boy whose father was a merchant or craftsman of any type would have grown up hearing how business was conducted.

Given the somewhat autocratic nature of the period, you might expect that the apprentice had little chance to be heard if he felt that he had a bad deal, but this was not so. It was in everyone's interest that the tuition was carried out competently and that the lodge or guild did not get a reputation for endorsing poor training. That could have had an effect not just on future employment and projects, but also on the longevity of the buildings. Although it is vague in its wording, the Statutes include the following clause:

If an apprentice considers that a master is not fulfilling his duty towards him in any respect, as he agreed to do, the apprentice may bring the matter before the corporation and the masters who reside in that area, for a full investigation.[4]

One of the issues that a guild might have investigated was whether or not an apprentice was being allowed to progress. Without doubt apprentices would have made friends among those being trained by other Master Masons and they will certainly have swapped details about each other's training. It must have been obvious at times that some students appeared to be leaping ahead while others lagged behind, not necessarily because they were not quick learners. Some Master Masons were more giving in sharing their skills than others, so the regulations covered that, too.

The thirteenth article of the *Regius Manuscript* says that the master who hires an apprentice 'should make everything known of the art whose slightest detail he has consented to reveal. The pupil will make this his store of knowledge.'[5] Interestingly, it goes on to say that the Master Mason is obliged not to offer an apprenticeship if he knows that he cannot offer a variety of work that would complete that information store. This implies that the master should consider the intricacies of known work in the pipeline and be confident that he could introduce the aspiring mason to a complete range of techniques. If he could not do so, then plainly he was wasting the youngster's time, which would have been a major source of friction. No record has yet been found of an example of that happening, but the fact that it was included in the regulations suggests that it must have occurred at least once and that the apprentice had been vociferous enough in his protests for the authorities to want to ward off future complaints.

Having said that, the apprentice complained at his peril. If the authorities found against him, he could find himself ousted from the community and no longer an apprentice, no matter how well he had been doing in his training. Apprentices' positions were further protected by the rule that no one was allowed to incite them to leave (although the fact that the clause below was included at all indicates that there had been cases where the hapless trainee had been leaned on and made so miserable that he felt he had little choice but to abandon his training):

No master or journeyman may incite an apprentice who is tied to him to leave, or dismiss him, or take on another one, who has come from somewhere else, unless he first obtained the authorisation of his master in such a way that he may leave him without any grievance. But if this were to happen, the master responsible must be brought before the corporation and be punished.[6]

Punishment was part of the training process but, perhaps surprisingly, that was moderated, too. In those days, when corporal punishment was the norm, we find that in sixteenth-century Germany, beatings were banned unless the master had prior knowledge of it and had given his consent. This is something also found in the Rule of St Benedict, written c.500, which became the most widely used set of operating procedures for monasteries. While on the one hand, boys could be beaten if they were deemed to be too dim to understand what a serious punishment excommunication was, no one was allowed to strike anyone unless the abbot had given his authority. The monks were working in a very different environment, but the principle remains the same: no physical violence unless permission had been granted. This allowed for consideration of the situation and also for tempers to go off the boil before action was taken, but they were also anxious that people kept their feet on the ground no matter how high up the ladder they climbed. One of the clauses in the *Regius Manuscript* exhorts masters to remember how it was when they were mere apprentices, dumb with fear at the outset of their training.

Another safeguard was that apprentices could not be given responsibility above their level of competence. One of the clauses says that the Master Mason could not make his apprentice a supervisor, indeed he could not take on a supervisory role at all until he had not only graduated, but completed his wander years. All of these things helped the apprentice uphold his part of the bargain for which he was often paying good money in any case. The Ratisbon Statutes of St Michael refer to money to be paid to the corporation when a Master Mason first took on an apprentice, and a further payment on completion of training:

> When the apprentice is declared free, he will be asked to pay one florin, but never more, to be spent on drinks for those who are present and who witness the granting of that freedom.[7]

While this was not a graduation ceremony as such, there was ritual that marked the achievement of both master and apprentice. Qualifications and freedom could only be granted formally; it was not acceptable for the master simply to say that the training was complete and that the apprentice was now a freeman. There would have been little benefit to either party if that had been allowed. The ceremony marked a definite end to the training period, which was made public, perhaps so that jobs might start being offered to the most promising newly qualified craftsmen. The master was now at liberty to take on another apprentice, so his success in bringing a youngster to maturity would be widely known and, if the training was reckoned to have gone smoothly, perhaps he was equally besieged by those hoping to be his next student. At the outset of the apprenticeship, a card was prepared that had a unique pattern running down the centre of it. The card (or piece of wood) was cut into two along this pattern, half of it being stored in the lodge and the other with the security payment. The idea was that this indented card was peculiar to a particular apprentice and acted as a way of recording that he had officially been taken on as a trainee and had paid his fee (from which we get the term 'indenture'). The system seems to have begun as a contract between apprentice and master, but developed so that it was eventually organized by the guilds. Naturally enough, it did happen that sometimes an apprentice wanted to cut short his training or perhaps to change master as we have seen with Antonello da Messina's student. This was also covered by the regulations, which were nothing if not thorough.

If it should happen that an apprentice were to leave his master during his years of apprenticeship without a valid reason, and thus did not serve his full term, no master was allowed to employ such an apprentice and no one must remain near him or have any company with him, until he had served his years in an honourable way with the master whom he left. He must have written confirmation from his master that he had done so. No apprentice was allowed to ask for a refund of the security payment unless he was getting married with his master's consent, or unless he had other valid reasons that compelled him or his master to make this request. This may only be granted if the brotherhood had been informed and following a judgement made by the stone-cutters.

Human nature being what it is, it was also natural that occasionally masters or apprentices disobeyed the laid-down articles. If that happened they would be summoned to appear before the corporation, where they would be given punishments in accordance with the oath and the commitments that everybody had taken in respect of the guild. This was usually in the form of a fine, but it could ultimately be expulsion from the profession, or just being shunned for a while.

> But if anyone scorns the punishment or summons without a valid reason and does not appear, the punishment will still be imposed, even if he is not present. If he does not respect it at all, he will not be permitted to do anything at all.[8]

That last sentence is the important one because it neatly took away any chance of employment, no matter how expert the craftsman might have been. Even so, the precise point at which the apprentice might expect to be admitted to the guild is not clear cut.

Another clause stipulated that every master who had apprentices 'must loyally encourage and invite each one who has completed the five-year term to become a brother',[9] which suggests that some did not want to. It is hard to know why they would not want to be a formal part of the guild because they became the controlling bodies; from them future employment, recommendation and protection flowed. Perhaps they were altogether too controlling for some of the Master Masons but, as we have occasionally seen already, most Master Masons were not exactly shrinking violets. Some of the candidates must have signed up to be part of the brotherhood in the same spirit that some join unions today: it is simply easier in the workplace to go along with it.

Uncertainty arises because it is also not clear at what point the lodge became the headquarters of the guild, and this varied in different regions. Much must have depended on the strength of will of individual Master Masons and how interested they were in having a corporate role. Certainly, in Germany the regulations show that freedom came with admittance to the guild, at which time certain promises were made by the new guild-member. These promises were made on oath, which was taken extremely seriously in the Middle Ages; breaking an oath could mean losing everything, even the right to communicate with other masons. One of the promises a new guild-member had to agree to was not to write down anything about the guild, which may account for the shortage of training manuals. He would also vow to obey the guild in all things and to accept its judgement, not to weaken the guild but to strengthen it and this would include not employing non-guild-members ('No one must cut stones next to him who is not a regular member of the corporation').[10] Intriguingly, these rules would also insist that the new guild-member promise never to reveal to anyone the masonic greeting or handshake. Links (or rather the tenuous nature of them) with modern Freemasons have already been mentioned, but this is said to be where the idea of an idiosyncratic greeting known only to its members began. The distinctive handshake would have acted as an additional identification check, but it was hardly foolproof. It is not revealed which party would offer his hand first in this way, but possibly it was the junior one, who wished to establish his credentials. Had it been the other way around, then presumably all sorts of uninitiated people might have learned how to do it. It also assumes that every single graduating apprentice kept his word, never showing off to his mates.

This was a closely knit, tightly run world that took its authority from God: the most prestigious buildings the Master Masons constructed were houses of God so, they could argue, theirs was the prime profession. They were, after all, building Heaven on earth or reconstructing Solomon's temple. They may well have felt themselves to be morally superior to other folk, which might account for the slightly endless list of strictures concerning the proper and honourable behaviour expected at all levels, not just the hapless apprentice. The morals of both trainee and master were rigorously examined, as were those of journeymen or companions, possibly even more so since they were away from the unblinking gaze of their mother-lodge. On this, the Statutes have much to say:

> No craftsman or master shall be admitted into the guild if he does not receive the Holy Sacrament once a year, or does not respect Christian discipline, or if he wastes his wages by gambling. But if anyone who was inadvertently admitted into the guild does not respect the above-mentioned principles, no master shall have any relations with him.

> The master must not hire any journeyman who is leading a dissolute life or who is living with a concubine, or who does not go to Confession once a year, and not receive Holy Communion, or who wastes his salary by gambling.

> No craftsman or master shall live openly with a concubine. If, however, such a person does not wish to stop living in this state, no travelling journeyman or stone-cutter shall remain in his service or have any relations with him.
> No master or craftsman shall employ a journeyman who is living in a state of adultery with a woman or who is living a dishonourable life with women, or who does not go to Holy Communion in accordance with Christian teaching, or who is foolish enough to gamble away his clothes.[11]

It would be interesting to know how seriously some of this was taken. It may well have been that even some of those who administered the regulations were guilty of dissolute behaviour, although the records are silent on that point. It is telling that there are so many strictures laid down and also that the Statutes of St Michael remind the Master Mason that even he shall behave in accordance with the rules and should take them as a guide. This should not have been hard since the basic principles of the guilds were of fraternity, honesty, oath-keeping, good morals, professional training, prayer and keeping the knowledge of the trade and guild secret – this last being a recurring theme.

The wander years

Having qualified, the former apprentice now became a journeyman, and was expected to go on the road on what were sometimes known as his wander years. This was where he really started to learn his trade, as one does when one has had plenty of supervision, but perhaps little chance to make one's own decisions. He need not have feared: the Statutes of Ratisbon, among others, had much guidance for him and, of course, plenty of safeguards, too. It may seem strange that, even though the now former apprentices were expected to go out to make their own way in the world, they also had to have permission to travel, but that was a widespread rule in medieval society. They had to bid farewell to their one-time masters and hostelry 'in such a way as not to owe anyone anything' and not to give anyone a reason to take issue with them later on. This was common practice for anyone going on a journey in the Middle Ages: travel could be fraught with danger so it was simple good sense to ensure sure that one's affairs were in order before setting off. There is evidence of this from the world of pilgrimage, not least the eccentric Margery Kempe of [King's] Lynn, who asked her parish priest to announce her departure for the Holy Land, saying that members of the congregation should speak up then if they thought that she owed them anything.[12]

Getting the job was the first challenge, even in days when there was an unprecedented amount of building happening right across Europe. A journeyman's best chance and recommendation was the name of the Master Mason who had trained him. If he had been apprenticed to one of the well-known names, such as the Ramseys or perhaps to Henry Yvele or Stephen Lote, finding work would be easy. Inevitably, it would have been getting that first position on a prestigious building site that shaped the rest of his career; many will have had to settle for work on ordinary parish churches, lesser secular projects or even – Heaven forbid! – constructing latrines, as happened to Geoffrey of Carlton in the fourteenth century. A journeyman would want to have trusted Master Masons standing surety for him. We can see from the documentary evidence how important this was for all parties because this was often recorded. For example, Alexander Abingdon guaranteed William de Hoo's work c.1300. Abingdon was a Master Mason who had begun his career as an imager (a sculptor and painter)

There appears to have been a rigorous application procedure to work on the cathedrals, but that may have been more in the regulations than in practice. There was certainly some enquiry into background, both in terms of morals and tuition. The potential employer was warned never to accept a pot of wine from a candidate (by which they meant a bribe); everything had to be above criticism, especially if anything went wrong with the building in the future. Nor could a candidate be offered a post that was already filled by someone else, although this must have happened if the journeyman was known to be a man of more skill or diligence. If he was still a journeyman, as opposed to a fully fledged mason or carver seeking employment, then the applicant would be watched closely to make sure that he did not speak out of turn, was not disruptive, did not go out carousing on the town and, indeed, did not even go out to eat without permission. That last stipulation may have been an attempt to control rowdy behaviour among the young men because the 1370 Rules for the Conduct of the Masons at York Minster state:

And in time of meat, at noon, they shall, at no time of the year, stray from
the lodge nor from the work aforesaid, more than the space of the time of an
hour, and after noon they may drink in the lodge.[13]

In case anyone was in any doubt, the Statutes were to be read to every journeyman once a year. The reading of the Statutes would have been a formal occasion,
just as novices and monks had to have the Rule of St Benedict read aloud to
them at various times. It will have arisen from ancient times when declarations
and instructions were delivered verbally to allow for those who could not read.
It was quicker and it was seen to have been done. At a time when literacy was
not widespread and books were expensive, this helped to ensure that everyone
knew what they were supposed to know. We cannot state categorically whether
apprentices, journeymen and masters could or could not read; they had to learn
geometry (which does not necessarily mean that they could read), draw up plans,
order materials and negotiate contracts – all of which could be done without
someone being literate. Of course, it would have been much more difficult for
someone if they were not literate and, although we do not know, it seems likely
that reading and writing must have been part of their training. Many were literate, as is evident by the surviving wills, plans and letters, and John Harvey points
out that Hugh Herland, William Vertue, James Nedham, Robert Carow and
Robert Lynsted were all employed as administrators as well as Master Masons.
Some of Henry Yvele's surviving documents show him working in French when
he had the administrative post of Warden of London Bridge in the later fourteenth century.[14]

Inevitably, the journeyman had to promise the corporation or guild to
abide by the rules, no matter how stringent. If anyone refused, he could not be
employed by the particular Master Mason to whose site he had applied or 'anyone
who may hear about it' later, something of a catch-all, which would have helped
to concentrate their minds.

In the journeyman's favour, the Master Mason was also advised never to
expect or ask him to do more than his skill level allowed; everything must be
within his capability (even though that might not have been fully appreciated
until some time had passed). The journeyman must be paid a salary that reflected
the local cost of living and he must never be paid less than he had been paid
before. All of that suggests that the Master Masons who stood as guarantors knew
their candidates well, and were either present at the interview or had written at
some length (although, if that is the case, the references have not survived). The
regulations have much to say about how a journeyman must not speak ill of his
employer, unless that employer had not followed the rules to the letter himself, in
which case it seems to have been acceptable for everyone to denounce him. That

same employer had to have a good reason for discharging a man from his service at any time other than on a Saturday (ideally giving him notice before noon) or on the evening of a payday. The idea behind this was that the journeyman should have the best chance of surviving financially until he found other work; he would be able to travel on the Sunday, as no one was supposed to work on that day, and be ready to present himself at a new site on Monday morning if there was one within about ten miles. Should the journeyman or any worker find himself in this situation, then he should accept it without debate or reproach, which sounds good in theory; what really happened may have been best left unrecorded.

There were cases when a journeyman decided to move on at short notice, presumably because of a personality clash or a better offer and, if that happened, he usually had to pay a fine to the guild. Leaders of guilds were anxious to show themselves to be what we would term good 'man-managers'; throwing a man out before he had a chance to be paid everything due to him would hardly have ticked that box. An exception to this is the nicely worded clause that no one should abstain from working on Mondays, the implication being that the weekend had rather caught up with them and they were not capable (in which case, a journey-man could be dismissed immediately). Master Masons themselves had all sworn at some point that they would pay rightful wages on time, which was intended to be further protection for employees. Life being what it is, not every Master Mason paid up in good order, so there are records of a few being prosecuted. This suggests that workers were not necessarily afraid to have their voices heard. The details are sketchy but these are not likely to have been peasant manual workers, rather they will have been qualified men who knew the worth of their hire.

Community chest

If, however, the journeyman (or any worker) had to be let go because of ill health, then help was, theoretically, at hand. The Statutes of Ratisbon and St Michael both refer more than once to a chest, which may well have been (and prob-ably was) a physical container not dissimilar to the community chests still found in numerous parish churches. These wooden chests normally had at least three different locks, the idea being that the keys would be held by three un-related individuals, all of whom would have to be present at its unlocking. Plainly, there is a way around this, but it was designed to prevent pilfering.

In the context of a guild, the chest was linked to the keeping of a book in which names, transactions, subscriptions, donations, payments, loans, etc, were recorded. In some places the book seems to have been used for rough minutes, records of events and even occasional records of punishment. It was multifunctional in that oaths could be sworn on it, even though this was still a time when it was more likely that a holy book or relic would be used for this purpose. Inevitably, there were rules about who could keep such a book and/or have access it (usually only the Master Masons and the guild hierarchy). No one was allowed to make a copy of the book for their own purpose, but such a book reveals to us more of the organization as well as the concerns of the guilds and lodges. To start with everyone had to pay a subscription, sometimes weekly, and that applied to the highest and the lowest on site including apprentices. In fifteenth-century Germany, the amount was placed at one pfennig a week for journeymen, which was dropped into the chest, presumably in front of witnesses, on payday. The money was handed over to the corporation or guild once a year. This accounts for the substantial size of some of the chests, which must have been quite an effort to move. But, before it went anywhere, the money was counted in front of each district supervisor to prevent fraud. The funds were for the good of all with no apparent proportions given according to rank or status. If a member of a guild deserved help, then they were given it, the word 'deserve' being crucial:

> If a master or journeyman [or apprentice or working craftsman] falls ill or if a member of the guild who has spent his time as a law-abiding mason falls ill, and therefore cannot ensure his livelihood or obtain the necessities of life, the master who has the chest and who is responsible for it must help him and assist him with a loan from the chest, if nothing else can be done and until such a time as he has recovered his health. The said member will have to promise to return the money borrowed from the chest; but if he should die in the course of that illness, a portion of his clothing or other belongings equivalent to the sum he was loaned will be confiscated, if that proves possible, out of the total assets left after his death.[15]

Part of the Master Mason's duty was to arrange to have Mass said as soon as he heard of a worker's death; everyone was expected to attend the funeral and to make a small donation. The chest was also at the disposal of those who had fallen on hard times, not necessarily connected with health, and for those who fell foul of the law, although the wording slightly implies that that was for those who were of good standing, not any old riff-raff who had got themselves into trouble. Naturally, the chest also covered legitimate and reasonable expenses incurred by anyone on the site, so long as they had been approved in advance. All of this shows a sophisticated level of management that had the good of the individuals, the guild and, most importantly, that of the building being constructed at heart.

The master was, of course, to provide housing for the journeymen and the numerous workers on site. Remarkably few would have been local men with a home to go to and it seems that the lodge building was not often routinely used for sleeping. Temporary accommodation blocks were erected for the workforce if billets were not available in town and they were not likely to have been of the highest quality, but were probably more like shelters for eating and sleeping. As the building site generally closed down for at least two months over winter, the Master Mason may not have wanted to spend money making them especially weather-proof. Where the journeymen went in that time is not recorded; the majority would have been too far from home to have returned, so they must have shifted for themselves as best they could until the site reopened on 2 February (the feast of Candelmas, also known as the Purification of Our Lady).

On the plus side, meals were provided. We have already seen that journeymen had to get permission to dine out, so to speak, and it seems that catering was largely done centrally in a refectory system, copied from monasteries. Apprentices will have lived and dined with their masters. The daily fare in the refectory is unknown, but probably there was not often meat because that is recorded as being a bonus on particular religious feast days. Other bonuses took the form of drink in time-honoured fashion. At Leicester, for example, in 1314 and 1325, and at London Bridge from 1404–18, beer was supplied on Ash Wednesday. In the accounts the sum of 3s 4d is noted down as having been paid for 20 gallons (90l) for those working on the bridge in general and for the masons and carpenters.[16] The *Regius Manuscript* speaks of communal dining in the great refectory hall at which the journeymen must be treated as though they are the brothers and sisters of the intendant (the master who is in charge of the hall for the week). It seems that there was a weekly rota for feeding the workforce and no Master Mason could expect to wriggle out of his turn, which must have been expensive. They were urged to make sure that they paid all the suppliers in full and on time and that they should publish the cost of it all. This is curious, but the most likely explanation is so that no master bought cheaper food or provided lesser rations than his colleagues. Everyone must be very well fed and the 'horrible name of miser' must never be levelled at a Master Mason when it was his turn to be host; ill feeling would be the inevitable outcome and he would be shamed.

In all of the three sets of regulations that have formed the basis of this section, feeding the troops has only one entry, whereas good, upstanding behaviour has several, as does the subject of secrecy. For example:

His master's counsel he keep and close,
And his fellows by his good purpose.
The privaties of the chamber tell he no man,
Nor in the lodge whatsoever they have done

... Whatever is said in the lodge is the secret of every mason: the master's words, advice, praise, criticism too, rules of the art, debates in a large or small room should never, by chance, leave the circle of its members. Observe the law of silence, by your honour, do not waver.[17]

The apprentice was thought to be learning the particular secrets of his master. Every Master Mason had a specific knowledge of his trade that he would share only with his successors. This was not unreasonable because it was through this knowledge that the master obtained his more prestigious contracts, even if it turned out that he knew no more or less than any other skilled Master Mason. When the master was enclosed with his apprentice he was said to be 'revealing his arts', as though such knowledge were some sort of magic. It was certainly the case that a man who had been apprenticed to a famous Master Mason would have a good chance of employment, possibly because his new employers were hoping (if not demanding) that they be let into the perceived secrets. It was not just apprentices who might be given access to the inner sanctum of knowledge. The Master Mason was advised to take a less-than-competent journeyman under his wing for on-the-job training if such a worker was on the same building site:

He should teach him to make the best use of his tools, to cut the stone better and to produce better blocks. Show him how to go about it, be charitable in your advice ... if he knows how to understand you, reveal your art to him.[18]

The mysteries of the guilds

The workings of guilds and individual lodges were deeply entwined in secrecy: this may seem reasonable as businesses to this day need to keep their own counsel, but it seems more likely that much of this secrecy was hyped up to increase the mystery. It would have been all but impossible to keep some of the secrets – after all, the physical act of building took place in the full view of everyone so nothing would have stayed secret for long. Indeed, some of the building techniques would have come under close scrutiny when considering building control and safety. Experienced Master Masons participating in Expertises needed to know precisely how part of a structure had been created so that they could be sure that a structure as a whole was strong enough. The idea of close confidentiality worked both ways. One clause ran:

Nothing shall be kept secret from someone who has been accepted and declared free, but everything that should be said or read to him will be transmitted to him, so that nobody can complain or use the excuse that if he had known this earlier, he would not have joined the corporation.[19]

It sounds almost as if these words had been written with a disgruntled mason in mind. It was a strange world to negotiate where a Master Mason had to balance the importance of not revealing what he was doing, or what his latest ideas might be, against the need to be open with his workers about what he wanted them to do.

It was also true that every building under construction, or even being planned, benefited from an exchange of ideas from all over Europe and beyond. Those young men on their wander years were of immense value to new masters and, of course, they wanted to share and show off their skill. Indeed, if they had mastered some new technique they would have been fools not to demonstrate it if they wanted employment. It is not hard to see how patterns would travel, one church influencing the design and/or decoration of another, and this will be considered in the next chapter.

It must come as no surprise to learn that the masters, too, were regulated, and not only with strictures that they must be honest and generally work so well that they deserved both their rest and pay. Would such a stricture really need to be written down? Apparently so, as was the (perhaps optimistic) suggestion that masters should always make others seem better than themselves and generally rejoice if someone succeeded where they had not. Shoddy work, on the other hand, ultimately led to expulsion from the guild, especially if this had resulted from idleness. Master Masons were reminded that they owed their presence to the greater group and that they should, in effect, see themselves as individual cogs in one enormous machine, a machine that involved the whole community and which had been functioning since time immemorial. The Master Mason was defined as being one of those members who were able to put up magnificent buildings for which they had received authorization, that last part always reminding them that they should never act on their own. If, for some reason, the Master Mason was not competent, no journeyman was to assist him – although how the journeyman was to judge his potential boss's ability is not revealed, especially as no two Master Masons were allowed to run the same project unless there was a rush to get it finished. Certainly, the authorities did not want a clash of personalities on site or, possibly worse, arguments leading to delays over the final design. Even if a Master Mason died, his replacement was obliged to work to the previously approved plan and neither modify it or bring its efficacy into doubt.

There were other, smaller rules by which Master Masons had to abide. Some of these were in place to protect the labour force. For example, if workers had been taken on to dig foundations or build a wall, they could not then be redeployed to do more intricate work that might be beyond their expertise, even if they insisted that they were skilled enough. Even the master's pay was supervised. This was usually negotiated directly with the Workshop Committee or patron but, if the money was not forthcoming for any length of time, then the Master Mason was not allowed to charge interest and nor could he do so if he had made a loan to the City or a corporation, perhaps for materials. Presumably, this was to prevent usury, which was a crime condemned more than once in the Bible.[20]

Of course, there were sometimes complaints and squabbles, and there was guidance available to deal with these, too. This was generally along the lines of advising a Master Mason not to take it on himself to resolve an issue on site without hearing both sides of the argument, and the plaintiff was not permitted to take the issue elsewhere. It was also made clear that any conflict should be resolved strictly outside working hours. Everything came down to the oaths that all had sworn publicly and which were sacrosanct. Every practising mason, journeyman and apprentice swore oaths 'with great respect to his Lord, to be faithful to all aspects of the traditions, rules and Law ... in the name of the king':

> His actions must always be guided by the need to conform to every one of its points. He will not be tempted to disobey any of them or to break the pact that ties him to all the free masons by the oath that everyone swears before his peers. All are required to examine their skills strictly. Whoever refuses or fails to do his duty shall be brought before the Assembly.[21]

Breaking any part of such an oath could be punished severely. It was laid down that the governing bodies should examine each case very carefully and that no one should prejudge the issue: the accused was held to be innocent until proved guilty. After whatever was deemed to have been a fair consultation, the outcome was announced publicly and, if it went against the mason and he acknowledged that he had been at fault, he was given the chance to amend his ways. If he refused to change his conduct, he was, inevitably, expelled from the guild and, indeed, from the profession. It is hard to imagine that anyone would risk jeopardizing his whole future, especially after such a long and exacting training, but this rule is found in more than one of the Statutes, so it must have been thought to be a real possibility. The offender should be held in contempt by everyone for his treachery for he had wilfully opted to break the tie that bound him to his great profession. Not only was he expressly forbidden ever to practise again, but he put himself in danger of being thrown into prison, with all his savings and assets seized for the duration of the king's pleasure.

Regulations are fine when all agree to abide by them, but, as we have seen in Chapter 4, that was far from the case. Nevertheless, Master Masons seem to have managed perfectly well and remarkably few ended up being barred from their profession. As is often the way, these regulations will no doubt have been observed when ceremonial activities required, but, back on the building site, individuals will just have got on with the job in hand. What will have been taken seriously were the rules concerning training apprentices and journeymen because those people were still needing a specific form of support. That they also needed a firm hand at times also comes across clearly. Even so, many of the apprentices grew up to become some of the greatest and most innovative Master Masons. They had to be charismatic, intelligent and strong-minded, so discipline, as well as training, was the order of the day until they reached maturity.

CHAPTER SEVEN

IDEAS AND PEOPLE TRAVELLING

Ideas have no boundaries

THE PROOF THAT more lodge secrets were spilled than kept is in the design and decoration of numerous buildings. Inspiration was taken from many sources, and ideas travelled either with craftsmen or pattern books, sometimes both. Often, it is easier to see this in parish churches than in cathedrals, although the principle was the same.

At some point in the fourteenth century, someone in the west and middle of England dreamed up the idea of building a hexagonal porch. In medieval iconography, the number six indicated completion because there were six days of Creation, yet a six-sided porch is not the most obvious or practical feature to create, and could look cumbersome from the outside if not designed very carefully. A porch was used for parish meetings, marriages and all manner of administrative functions. A hexagon would give a more pleasing shape than a rectangle and would make it easier to generate discussion. In northern Europe it became common to add above the porch a room, which was used for secular business. Such a room probably developed because of the climate. Later on, marriages and baptisms gradually moved inside the main church so the porch became less important. Evidently, this innovative hexagonal shape did not catch on because only three (albeit high-status churches) feature such a porch: St Mary Redcliffe at Bristol, St Laurence's at Ludlow and St Mary's in Chipping Norton. At some point, someone must have travelled from one of these churches, seen its unusually shaped porch, and carried the idea to the other two churches.

Ideas came from everywhere. The notion that the Gothic style itself might have come from the Holy Land was discussed in Chapter 1. It is possible that a Master Mason called Lalys arrived in south Wales from the Middle East, and we know that some Master Masons went on pilgrimage to the Holy Land. The Englishman who became known as Jean Langlois, and who was in charge of the finances at Saint-Urbain's in Troyes, travelled to Jerusalem in 1267. There is nothing to indicate that he returned, with or without ideas. John Harry disappeared from the accounts of Exeter cathedral from 1430 to 1432 because he had gone on pilgrimage to an unspecified destination, although the length of time suggests that his destination was a major shrine and possibly also somewhere in the Holy Land. John Palterton became a pilgrim in 1355, returning on 6 Feb 1356/7, his place having been taken by John East in the meantime. Taking time out from work in this way had its risks for a Master Mason (although not in terms of salvation) because who knew when he might return or how good his stand-in might prove to be. Palterton probably did not overly care, given that it was he who had complained so vehemently about the late delivery of his robes that he had refused them when they did arrive.

We know by name some of the Master Masons who travelled overseas for work purposes, as well as the names of some of who came to Britain from other lands. We also know where they went because their names appear in accounts and contracts. From these, it appears that the early British Master Masons were travelling to some fairly diverse places. One of these is William the Englishman, who built a church at St Jean d'Acre, in Palestine, in or about 1192. It is possible, though unproven, that he is the same man who took over from William of Sens in the rebuilding of Canterbury cathedral after his disastrous accident in 1177. The William who took over from William of Sens was probably known as 'the Englishman' to distinguish him from the French William, who was only the second recorded Master Mason migrating to our shores (the first having been Lalys). Although William of Sens is usually said to have made the Gothic style fashionable in England, being a classic example of ideas travelling, the Englishman surpassed him because the Trinity Chapel and Corona of Canterbury cathedral are technically and stylistically more advanced than the work that preceded Sens' accident.

Surviving documents reveal a great amount of travel taking place in support of a building, even for just a small part of it. When the doors for the baptistery of Florence cathedral were being considered in 1329, the records show that Piero di Jacopo was sent to Pisa to look at doors there and make sketches of them. After visiting Pisa, he travelled to Venice to find someone capable of doing the work. After that, several names are listed of people who worked on the doors, not all of them from Florence, so they were costly items not only in terms of materials used, but in travelling expenses, too.[1]

The earliest English Master Mason recorded working oversees was Walter Coorland, who was said to have been sent to Poitiers by Queen Emma (then married to Cnut) in 1025. Coorland remained at Poitiers for the best part of twenty-five years. His task was to reconstruct the gigantic church, now dedicated to St Hilary, where the saint and his daughter, St Abre, are entombed. St Hilary had been born in Poitiers at the end of the third century and became not only its bishop, but a Doctor of the Church and a highly influential figure in the medieval church. The Byzantine influences in this extraordinary building are plain for any visitor to see, and how an English architect came to have those in his repertoire may be clearer to understand than expected. It should still be noted that one of the many problems with trying to trace ideas travelling is that they quickly become multi-layered; one style influenced another and was later changed by someone else, who may have approached it in a different way, often according to the space and material available. It can be like trying to retell a dream that escapes the teller as soon as they start to describe it. We know nothing of Walter Coorland before or after this commission, so we have no idea where else he might have travelled and what else he might have observed that impressed him, but the very fact that he was working in Poitiers is important, as we shall see.

Byzantine influences

The origins of Byzantine influence on Europe art and architecture are comparatively straightforward because of the links between Venice and Constantinople. The former had been established as a trading post on the gloriously precise date of 25 March 421 at noon (this was the founding of the church of St James; St Mark would not come to prominence in Venice for another few hundred years). Venice is only about eighty miles north of Ravenna, another important trading post and power base from the fifth century onwards. Both cities' geographical positions gave access to what we now call Greece and Turkey, which meant also to the Black Sea and the Silk Road beyond. As an eight-year-old child, Theodoric, who was to rule Italy from Ravenna from 493 to 526, had been sent as a hostage to Constantinople and held there for a decade. While there he was educated as any young nobleman should be, and he must have absorbed ideas about using decoration as an expression of power. We see this manifested not simply in his mausoleum on the edge of Ravenna, but in the church of Sant'Apollinare Nuovo, which he had built for Arian practice, and which he lined with mosaics. What happened in Ravenna affected what happened in Rome because Theodoric saw the repair and re-ornamentation of Rome as an important goal of his reign. A century and a half later, that meant that those ideas were taken to Northumbria, carried there by Benedict Biscop, who visited Rome five, possibly six times, in the second half of the seventh century, and who founded the monasteries at

Monkwearmouth and Jarrow in the north-east of England. The Byzantine look of the figures on the early eighth-century Franks Casket, which was made in Northumbria, helps to illustrate this point, as do the Evangelist portraits in the Lindisfarne Gospels made c.700 off England's north-east coast.[2] One could also cite the imagery etched on St Cuthbert's coffin (died 687), currently on display in Durham Cathedral's treasury, which has further echoes in the Book of Kells made on Iona c.800.

Ravenna also yields the stunning mid-sixth-century church of San Vitale, equally beautifully lined with mosaics. It is of gigantic dimensions. Today it is a sizeable church; then it must have been all but impossible to comprehend. We have no idea who built it but they must have taken on celebrity status and there will have been great prestige attached to anyone who was involved with its construction. The interior is almost entirely covered with glittering mosaics and marble, gorgeous to behold and a glimpse of both heavenly paradise and earthly power. The attention to detail in its construction is fascinating; for instance, whereas the gold glass tesserae of the Sant'Apollinare Nuovo mosaics are laid flat with near-seamless joins to create an illusion of solid gold, those in San Vitale were deliberately placed at slight angles. This allows candle and other light to glance off them in different directions, adding to the dazzling effect from every viewpoint. The Byzantine emperor Justinian the Great commissioned it, having the design modelled closely on that of the Hagia Sophia in Constantinople (although he never went to Ravenna and, therefore, never saw it). Charlemagne, who ruled the Carolingian Empire from 768 to 814, visited Ravenna three times. On one of those visits, we must assume, he stood in San Vitale and decided that he wanted such a church at his static court in Aachen. The result is the basilica that stands to this day, an indisputable architectural link with the power bases of Ravenna and Constantinople with its tiers of thick-walled, soaring arches and mosaics set in a sea of gold. Pope Adrian I had authorized Charlemagne to take whatever he liked from Ravenna. It is not known precisely which pieces he had shipped north, but statues and columns seem likely. Just as the church of St Mark in Venice is adorned with marble revetments (and other things) taken from Constantinople, so the idea of slicing marble to reveal unusual patterns became popular in numerous European churches, whether using real marble or painted copies as at Saint Savin in France.

The Herefordshire School of Architecture and Sculpture

Another traceable handing-on of ideas from mason to mason or patron to patron can be found in what has become known as the Herefordshire School of

Architecture and Sculpture. It is characterized by, again, somewhat Byzantine-looking faces with almond-shaped, often bulbous eyes with deeply drilled pupils. If angels are featured, they tend to look a little glum and stolid; draperies are not so much fluttering as ribbed. There are often strangely contorted figures, hybrid and zoomorphic beasts with smoothly carved faces. It is not uncommon to have a repeated group of such creatures on a frieze or tympanum, such as at Aulnay near Poitiers. The zenith of this style in England is found in the church at Kilpeck, in Herefordshire. Constructed in the 1140s, this church has become famous for its corbels, but there are examples to be found on the fonts at Eardisley, Stottesdon, the now-folly arches at Shobdon as well as at Elkstone, Hereford Cathedral, Leominster, Rowlstone, Pipe Aston, Ribbesford, Rock (all in the west of England) to name but a few. The manner of carving that is called the Herefordshire School in Britain can also be found to this day in central and western France and in north Spain.

The story of how the style came to the west and middle of England lies with a man called Oliver de Merlimond, patron and traveller.[3] He was recognized at the time as being a man of intellect and learning, not to mention sound administrative ability because Hugh de Mortimer appointed him as chief steward of all his considerable estates. As part of the package, he gave Merlimond the hamlet of Shobdon, where there was a chapel but no church at the time. Being a pious man, Merlimond decided to go on pilgrimage to one of the great European destinations, Santiago de Compostela. He delegated his managerial duties and set off. This pilgrimage is thought to have taken place between 1125 and 1130, so he cannot have been influenced by the Pórtico de la Gloria because, if the inscription carved on this building is correct, it was not erected until 1188, i.e. after Merlimond had visited. We have a few clues as to what Merlimond saw because a collage of unconnected carvings in different styles has been placed around a doorway at the back of the cathedral and one of these has the date 1103 incised on it. It is not very much to go on and, as is often the case, the huge church at Santiago was built to replace another, not to mention its restoration after a disastrous fire in 1117, and we cannot tell what was there before. We do know that Merlimond returned to England via Paris which, again, does not help in terms of ideas travelling unless he went by land and not by sea.

It seems strange if he did return by land because sailing, although hazardous, was by far the quickest, safest and cheapest way to travel, but the carvings we see at Kilpeck, Shobdon and other places in the so-called Herefordshire School group have their cousins in churches on a land-route home. One answer may well be that he brought a carver or even a Master Mason back with him, in which case it would not matter which route he took, but there is no record of this. In Spain, the oldest church with this type of carving is the seventh-century Quintanilla de las Viñas, which Merlimond is unlikely to have seen because it is sited twenty-five miles south-east of Burgos – a considerable detour even for the joy of seeing unidentifiable animals leaping and the stylized foliage carved inside roundels. Likewise, the marvellous church at Frómista is some fifty miles west of Burgos and over 100 miles south from the coast. This church was built in the eleventh century at the behest of the King of Navarre's widow, so it was high-status; its decoration would have influenced other new-builds in the area especially, it seems, the goggly-eyed beasts and corbel heads that might possibly have been inspired by asses. On his return journey, Merlimond may well have visited Frómista and Quintanilla de las Viñas, which route would also have given him the chance to visit major shrines at León and Burgos. But if Merlimond returned to England by land, the logical route would have been along the northern Spanish coast rather than to trek across to León, Burgos and through the Pyrenees. On the coastal (and more feasible) route lies Santa Maria del Naranco, near Oviedo, whose church was built in 848 as part of a palace complex; again, very high-status. This offers Ravenna-esque high arches and capitals bearing stylized creatures that would have fascinated the viewer. Just over 100 miles eastwards, Merlimond would have come to the eleventh-century Santillana del Mar. While the carvings are not exact copies of each other, the figures beside the Kilpeck door echo one on a Santillana cloister capital. They wear the same shaped hat, are both entangled in a basket-weave pattern of tendrils, and, on other Santillana carvings, there are figures that sport drapery that is more ribbed than fluttering. The layout of the images is different, but details and the overall ideas are very similar.

Other artistic parallels can be drawn in French churches of the eleventh and twelfth centuries. Aulnay has some similarities to the Herefordshire School, but that façade was not constructed until 1120–40, which would have been too late for Merlimond to have learned much from it. The basilica at Toulouse (consecrated 1096) is too far east, but the façade at Saint Jouin-de-Marnes (1095–1130) and most particularly on Notre Dame la Grande, in Poitiers, would have been on his route, and they tick the right stylistic boxes.

The starting point in Poitiers has to be the baptistery, which dates to about 360, making it likely to be the oldest surviving church in the West. It was reworked in the sixth and seventh centuries, and then again in the tenth, which is when it gained its present octagonal shape. It was also in the tenth century that the Catholic Church stopped baptizing by total immersion, so the big pool now revealed inside it was filled in and the building became a parish church. On the outside under the eaves are corbels adorned with smooth-faced, non-specific beasts, which may be asses or pigs or hybrids with bulging, almond-shaped eyes, just as are seen at Kilpeck and in Spain. These motifs have been taken up by the carver of Notre Dame la Grande's façade in the twelfth century who expanded on them to glorious effect. In this case many of them tell Old and New Testament stories, but mingled in with beasties and dragons. Some of the stranger creatures echo those at Aulnay, which was built at about the same time, although we cannot know who was copying whom or, more likely, if it was the same craftsman. What we cannot help but notice is that variations of the same designs appear on the front of the vast church of St Hilary, also in Poitiers, the church to which Queen Emma had dispatched Walter Coorland in 1025. Could it be that it was an Englishman who was instrumental in establishing the style that would become known as the Herefordshire School?

Scandinavian influences

It is also evident that features of the Herefordshire School of Architecture, and numerous other unattributed church designs and decorations, hail from Scandinavia. This is not surprising, given that Viking attacks were so violent that they were well recorded, so they are known to have reached not just Britain, but also the areas now known as France and Spain, and, indeed, beyond. The first recorded full-scale attack in Britain was on Lindisfarne in June 793, but there had been an incident on the south coast before that. The entry for the year 789 in the *Anglo-Saxon Chronicle* tells us that three ships of Northmen landed at Portland. The reeve went to meet them to find out what their business was, addressed them rather haughtily and was killed for his lack of diplomacy. Contemporary writings have a sense of shock because there had long been trade between Scandinavia and Britain, which helps to account for initial exchanges of ideas between those two regions.

    While we do not have much evidence of Scandinavian building, whether religious or secular, we do have numerous runes and picture stones, wealthy burial sites, jewellery and, as time went by, stave churches. Scandinavia did not convert to Christianity, generally speaking, until about the year 1000, and the oldest surviving stave church dates to roughly 1180–1250 at Borgund, in Norway. Inevitably, it has been much restored, but stave churches were solidly built, consisting of interlocking wooden boards that would have taken a very differently trained Master Mason to construct. The art forms in all the media mentioned are characterized by daintiness, surprisingly so for such an aggressive military force. James Graham-Campbell's book on the subject breaks it down into six main phases between the dates of *c.*775 and *c.*1125, so a significant amount of it is not Christian. While there are grotesque monsters (for example, in the Oseberg ship burial of 834), the figures are beautifully carved with much skilled intertwining of foliate tendrils. This is seen throughout Viking decoration, almost always with patterns covering every available inch of a surface. The delight in filling every space is something that happens in Gothic art; it is rare to see empty spaces on the front of a cathedral and, if we do, it can mean that there was paintwork that has been lost to us. The Scandinavians liked zoomorphic creatures, often spouting elegant tendrils. Equally, often the 'animal' depicted is all almond-shaped eyes and swirling joints, as on the door surround at Urnes, in Norway. Or it might be a nastily contorted beast with an excessive number of vice-like talons clawing at the edge of a brooch, a motif known as 'the gripping beast'. A rune stone found in St Paul's churchyard in London shows a running animal looking back over its shoulder. There is a freedom about it, despite it being entangled in delicate wisps of what may be plant life. Viking influences on church decoration can be found across Europe, whether it is the font at Hinton Parva, in Wiltshire, which has a similar animal looking back over its shoulder, or the gripping cat-beast on a capital in Melbourne, in Derbyshire.

Whatever the image, they were created with great skill, not least the treasures that were found in the ship burial at Sutton Hoo, near Woodbridge in Suffolk. The burial probably took place in 615, so it significantly pre-dates Viking attack, but falls into the time when, we assume, trading was comparatively peaceful. It is an extremely high-status grave, remarkable for the fact that all of the thousands of garnets buried there have been tested, and all are believed to have come from India. That aside, the workmanship of buckles, rings and shoulder bags is exquisite, many of these being decorated with granulation. This is a simple design that comprises two parallel lines with a row of fine dots running between them. It is a design that is seen endlessly in churches and cathedrals across Europe up to about the year 1200, when it began to go out of fashion.

Persian influences

Robert the Sculptor will have seen a good range of artistic and architectural styles when he accompanied Sir Geoffrey de Langley on a diplomatic mission to the Ilkhan of Persia in 1292. Very little is known about this intriguing expedition, which would have been the talk of all society. Trade routes were well established across the known world (knowledge of which was about to be expanded even further by Marco Polo), but this journey must have been the stuff of fantasies. Happily, some of the financial accounts have survived and we also know that Ghazan Khan maintained the friendly relations with western Europe that he had inherited from his father, Kaikhatu (great-nephew of Kublai Khan), so the destination may not have been as dangerous as it might sound, even though travelling there would have been fraught with difficulty.[4] Edward I accredited

Geoffrey de Langley and two esquires to the Persian Court probably as the result of an invitation from the Ilkhan to view his new bride. They will have travelled with numerous aides and assistants, some of whom will have been there to add prestige and culture to the party and perhaps to help present suitable gifts, and Robert the Sculptor was one of these. On his journey Sir Geoffrey bought silver plate, fur pelisses and carpets. The party travelled via Genoa to Trebizond and Tabriz, returning home with a leopard, which would have joined other exotic beasts at the medieval zoo at the Tower of London. Unfortunately, all we know of Robert is that he was with the party; whether he put his new artistic ideas into practice has not been revealed.

English Master Masons abroad

Some of the projects allocated to British Master Masons overseas were extremely high-status. When John of Gaunt's daughter, Philippa, left for her marriage to the Portuguese King John on 2 February 1387, she took with her a train of courtiers and cultured men, similar to Sir Geoffrey's to Persia, but even more impressive. How the Master Mason Stephen Stephenson was selected to be part of that entourage is not known, but he was quickly spotted by Affonso Domingues, the chief master of the new works at Batalha Abbey. This abbey is one of the largest and most important in Portugal, and is now a world cultural heritage building. We know that Stephenson's designs found favour because the English influences on the building are very strong, so much so that a tourist keeping an account of his travels on the internet said recently that, 'The nave was huge and reminded us of some of the large cathedrals in England.' Quite! The writer could also have mentioned the very English way of placing roof bosses in the cloisters, and some of the English curvilinear carvings, the arrival of which in Portugal coincides with that of Stephenson.

Having made this more than promising start, the records show that those English Master Masons who travelled later were predominantly concerned with warfare and were mainly working on fortifications, first at Harfleur and then at Calais. The period that we call the Hundred Years War[5] had begun in 1337 when Edward III made a claim to the French crown. The victory at Agincourt in 1415 saw the first group of Master Masons (John Clife, John of Colchestre, John Janyns, Thomas Matthew) crossing the English Channel to strengthen the walls of Harfleur, very little of which remain today. The work at Calais was more ongoing because Calais remained an English possession until 1558 when it was lost to the French and Queen Mary I of England famously said that when she was dead the name of Calais would be found engraved on her heart. The high status of the diplomatic talks at the Field of the Cloth of Gold between Henry VIII and François I of France in 1529 can be gauged by the quality of Master Masons sent to construct the ostentatiously lavish temporary accommodation. These included the three top masons at the time: Humphrey Coke, Thomas Stockton and William Vertue. Even the cloth of the tents was lavish, being made of silk interwoven with filaments of gold leaf, hence the name given to the site. With money being no object, Coke, Stockton and Vertue must have had a wonderful time, not just experimenting with new ideas in the full knowledge that their constructions would be tried out, then dismantled, with no harm done, but also doubtless visiting many French cathedrals and abbeys along the way. After all that, a tactless winning throw by François in an impromptu wrestling match between the two kings brought the whole thing to a conclusion with no treaty signed.

Foreign Master Masons active in England

While English Master Masons mainly worked on fortifications, the projects for foreign Master Masons coming to England were more varied. Their specialities ranged from mason and carpenter to tomb and brick-maker (John of Limoges and Baldwin the Dutchman), and they came from as far afield as the Holy Land, Germany, France, the Low Countries, Italy and Spain. Not all were involved in cathedral building. James of St George was a French castle builder for Edward I, working mostly in Wales. The ground plans of Caernarvon and Conway castles are so similar that he had evidently hit on a practical design and stuck to it. It is likely that the concentric plan at Harlech was also his. However, it was not his invention because the original idea arrived with returning Crusaders and is first seen in action in England at the Tower of London in 1274, some four years before St George's arrival. Likewise, Baldwin the brick-maker was a fortifications specialist. We might not think that a lot of skill is needed to make bricks, but this was a relatively new building form when he arrived in England in the fifteenth century, so he was brought across to advise on and supervise one of the earliest large-scale use of bricks at Tattershall Castle, in Lincolnshire.

Arguably one of the most influential Master Masons working in England was Henry de Reyns (of Rheims). Henry III had been in France in 1242–3 and had visited the newly consecrated cathedral at Rheims, the site of French coronations and royal occasions going back to Clovis' baptism at the end of the fifth century. This new cathedral must have been as breathtaking to view then as it still is today (*see* plate 12). To Henry III, this cathedral was the ultimate expression of everything spiritual, majestic, awesome and beautiful; nothing would do, but he must have a similar building near London. The obvious place for this was on the site of Edward the Confessor's great church at Westminster (*see* plate 13), and it must be created by the same craftsmen who had fashioned the church in Rheims. Again, it is not an exact copy, but the similarities between the two cathedrals are striking: if Henry III could have lived long enough to see it finished, he must surely have been satisfied.

Edward the Confessor's soon-to-be-replaced church had also been made along French (Norman) lines by Leofsi Duddason under Godwin Gretsyd's financial patronage. De Reyns was already in England working on Windsor castle and he was brought to Westminster, where he received his gown of office as Master Mason in December 1243. His first job was, dauntingly, to pull down the Saxon building and he began this on 6 July 1245, a good eighteen months after he had received the commission (clearly, this stage could not be rushed into, but required careful planning). He also had to hand over his Windsor project to a suitable replacement.

Under his direction, the new abbey grew rapidly, no doubt with the king chafing at the bit in the background. It could not be built fast enough for Henry III. In October 1251, he ordered his Master Mason to hasten the work by getting all the marble work raised during the winter so far as was possible without peril although, in reality, he probably did not particularly mind if the site was any more dangerous than usual. By 1252, they were ordering timber for the roof and stalls in an unprecedented race for the sky. Extraordinarily, Henry de Reyns was in charge of concurrent work on the Tower of London, but he disappeared from any record in 1253 when John of Gloucester took over the abbey work. He may have died, possibly of exhaustion and royal nagging. In 1260 there was a reference to one Hugh, 'son of the late Master Henry de Reyns', but there is no indication as to how late. De Reyns has been discussed with the general group of foreign Master Masons, but he might not actually have been one of them.

The medieval habit was to name a Master Mason after the last place with which they had been associated, which presents another problem when trying to track individuals because their names can change with every assignment. To make life slightly simpler they often seem to have retained the name relating to their greatest work, or Henry de Reyns would have become known as Henry of Windsor, and we might not have realized that he had come from Rheims before that. Instead, we would have wondered at what point was he in Rheims and in what capacity? There is no doubt at all, even to the most casual observer, that the churches at Rheims and Westminster come from the same stylistic stable. There is equally little doubt that what happened in artistic terms at Westminster enormously influenced later architects, so whoever brought the template across the Channel had a significant impact on what would happen later in Christian building in this country. The problem with laying this at Henry de Reyns' feet is that it is more than likely that he was already here, having arrived in 1239 to work at Windsor Castle. This was before Henry III conceived the idea of a new church. It has been suggested by W.R. Lethaby that on stylistic grounds the King's Chapel in Windsor Castle (1239–43) must have been designed by the architect of Westminster Abbey mainly because of the similarity of the moulding sections.[6] The romantic idea of Henry de Reyns' talent being spotted and his being snapped up by the English king does not stand up to scrutiny, and the most likely explanation of his name is that he spent his wander years in Rheims. That would make sense because he would have had the chance to study under French Master Masons; the design for Rheims did not, after all, come out of the blue, but was a bringing together of numerous structural and artistic experiments. He may well have been an Englishman who could have brought pattern books or templates with him when he returned to England. Again, the problem with tracking the changes in building styles is that there was rarely a new breakthrough moment and cathedrals took so long to build that ideas would seep across national boundaries well in advance of any unveiling.

Nicholas Dyminge made one of the Eleanor crosses as well as the queen's tomb in Lincoln cathedral, so he was in high royal favour. He was also known as 'de Reyns' and so was probably also from Rheims, or must have been there at some point. Many think he must have been a descendant or other relative of Henry de Reyns, in which case they were a talented family. It is debatable, as we have just discussed the difficulty of pinpointing origins.

A change to more naturalistic forms

Very occasionally, there is evidence that someone's work caused offence although we have no idea as to precisely why because no record of the item itself survives. One such case was that of a German carver called Tidemann, who was paid a hefty £23 to make a crucifix in 1306 for the church of St Mildred's, Poultry. This was a City of London church destroyed in 1450, rebuilt, then burned down in the Great Fire, rebuilt by Christopher Wren in 1670 and finally demolished in 1872. We are led to believe that crowds were flocking to see a crucifix that was making them excitable and unruly, so much so that it was confiscated by order of the bishop of London. It was handed back to Tidemann outside the City, either at early dawn or late in the evening, when it could be done without attracting undue attention and causing more scandal. He was also obliged to swear an oath that he would never make or sell such a thing again anywhere in London. 'Scandal' is an interesting word to use about a crucifix and, as so often, shows us how much more important Christianity was to the general public than now. The only thing we know about Tidemann's crucifix is that the arms were not of the 'true form', which might not sound very helpful if it were not that there were still set rules about how things should be portrayed. Iconographic rules were on the verge of changing, probably for reasons of artistic taste or technical advance, and perhaps as a reaction to the turbulent events of the twelfth and thirteenth centuries. There is a clue as to what might have happened because of a case brought in Spain in the late thirteenth century concerning so-called 'deformed' crucifixes. Bishop Tuy accused some Spanish craftsmen of:

> Painting or carving ill-shapen images of saints in order that by gazing on them the devotion of Christian folk may be turned to loathing. In derision and scorn of Christ's cross, they carve images of Our Lord with one foot laid over the other, so that both are pierced by a single nail, thus striving to annul men's faith in the Holy Cross, and the traditions of the sainted Fathers, by super-inducing these novelties.[7]

By the same token, a good 300 years later there was scandalized objection to Caravaggio's *St Matthew and the Angel* (*c.*1600). He made a point of introducing everyday, even coarse, people into his religious scenes, causing at least three pictures to be rejected by the ecclesiastical authorities who had commissioned them. St Matthew was shown with his legs crossed (a sign of either badness or power, or both in medieval iconography) and the bottom of his feet scuffed and filthy. Saints had lived in this world, but must not be seen to have been of it lest that detracted from their ability to intercede on our behalf, or in any way undermined their status.

It is certainly true that the earlier images of Christ crucified are stylized and formal: Christ's body usually echoes the perfect Cross shape with feet neatly nailed, side by side. In these images, it is rare for there to be any expression either of face or gesture or for there to be any implication of suffering or weight. The basic designs had first been laid down by the Second Council of Nicaea in 787, which helps explain the uniformity of composition across Europe. At this Council was discussed, among numerous other things, the first of the Ten Commandments:

> Thou shalt have no other gods before me. Thou shalt not make unto thee any graven image, or any likeness of any thing that is in Heaven above, or in the earth beneath, or that is in the water under the earth. Thou shalt not bow down thyself to them nor serve them: for I thy God am a jealous God, visiting the iniquity of the fathers upon the children unto the third and fourth generations of then that hate me.[8]

In addition to the reluctance to make images that might attract direct worship for themselves rather than be vehicles for devotion, there was an anxiety about reproducing God as He might look in human form. There was a great reticence about showing Christ's suffering and humiliation on the cross because that had been an instrument of execution for common criminals. It was only natural for what was still an emerging religion to show God as triumphant and almost non-committal about the pain of his particular death because God was more than a mere mortal: He transcended all. We can see something of the transition from stylized Christ figures to realistic ones in the wall paintings in St Albans abbey, where there is a set of five Crucifixion scenes. In the first, which probably dates to *c.*1215, the Nicaean rules have been followed, although there is a litheness about Christ's body that suggests a moving forward to a more natural style. The second Crucifixion was painted some twenty-five years later, but here the legs are twisted so that Christ's feet are on top of each other in a rather ungainly manner, and both are pierced through with a single nail. By the time the fifth Crucifixion was painted *c.*1275, all the rules had been abandoned and Christ is contorted with suffering, all too realistically. In her book on the wall paintings, Eileen Roberts remarks that crossed legs are a medieval symbol for the interruption of the normal flow of life. One could speculate that this is also why effigies on knights' tombs are often shown with their legs crossed at the ankles; it could indicate that the life was interrupted in that they were killed in action or perhaps died of wounds or disease encountered in foreign lands.

By the time the Black Death had swept through Europe in the 1340s, styles had changed dramatically, with Christ sometimes writhing in agony on a crudely fashioned cross, perhaps shaped more as a 'Y' than a 'T'. One of the effects of plague was a greater desire to consider the humanity of Christ. Portrayals of the Virgin and Child changed from a rigidly posed Mary staring blankly ahead with the Holy Infant nestled in an altar-like crib, to her interacting with her baby as any earthly mother would. This new way of showing figures first appeared at the beginning of the thirteenth century. Prior to this, it was considered inappropriate to make Mary and Child look too human, but at Rheims Mary is shown swaddling Jesus in the most natural manner, as has already been described. The infant is of realistic proportions, not the stiff dolly of earlier images, and Mary's movements are fluid.

Spreading ideas

Ideas will have been spread by junior masons and carvers on their wander years. As already discussed, it was normal for a young craftsman (known as a journeyman) to wander for two or three years, gaining experience on different building sites. Indeed, Rheims cathedral, which we have already seen to have been so influential, had numerous German stone-cutters working on it. When the Alsatian architect Niesenberger went to Milan to raise the dome of the cathedral in 1483, he took with him thirteen craftsmen from different countries.[9] 'I summoned the best painters I could find from different regions and reverently caused these [walls] to be repaired and becomingly painted with gold and precious colours', wrote Abbot Suger concerning his innovations at St-Denis.[10] The Master Mason who worked on the cathedral of Santiago de Compostela from 1078 was Bernard the Elder and he is reported to have arrived with about fifty stonemasons. A list of masons working at Windsor in 1365 shows a total of twelve from London, Norfolk and Gloucester, two each from Bedfordshire, Buckinghamshire, Lincolnshire and Somerset, one from Lancashire and one from Huntingdonshire. These were from the same country, but the list indicates that expertise would be recruited from a wide geographical area even within that one country. More remarkably, at Avignon, in France, a register has been found that details those who worked on the papal palace there between 1344 and 1345. There are nearly 300 names – and not all of them local, by any means. A major new project acted as a magnet to squads of workers, often foreigners or youngsters on their wander years. So many languages were spoken that the Master Mason's right-hand-man was his *parlier*, from the French 'to speak'. He was a polyglot whose role was to keep the building site working under the direction of the Master Mason, and without him, great confusion could ensue. The idea of the site mimicking the building of the Tower of Babel was popular and often illustrated in manuscripts.

Those working in the building trades were highly itinerant and they carried ideas with them in their heads or in pattern books. Books were portable, more so now than then, and they were a good way to spread artistic ideas. It is not at all uncommon to see a design that must have been lifted from a manuscript, such as on a chantry at Boxgrove Priory or, more dramatically, in the early twelfth-century murals that completely cover the walls of St Botolph's, Hardham in Sussex. Here the general form of the figures is seen not only in numerous manuscripts of the same period, but also in the Bayeux Tapestry.

The concept of pride was usually represented by a man falling off a horse, literally: pride before a fall. There is such an image in the Luttrell Psalter, made in about 1335-40, which has strong echoes in a misericord carving of the same subject in Ely cathedral made in 1338. It seems more likely that the wood-carver at Ely saw the image in the manuscript, or template for the manuscript, rather than the illustrator having seen the carving, but who knows? Perhaps both artists looked in the same pattern book? Similar interpretations of pride can be seen at Rheims, Chartres, Poitiers and Lincoln. At Ely there is another misericord showing the feast of Herod at which Salome danced and her mother told her to ask for the head of St John the Baptist. This bears an uncanny resemblance not just to a mural in the tiny, isolated church of St Hubert at Idsworth, in Hampshire, some counties away yet painted at almost the same date, but also to a roof boss in Norwich cathedral. All three of the images have Salome bending right over backwards executing a sword dance in a way that is more gymnastic than balletic. It is easy to see how a manuscript might have influenced a carver because it was the more portable; it could travel to him rather than he to it – but again, how can we be sure? However, when it comes to similarities between wall paintings and carvings, then it seems more likely that either a craftsman has travelled or that the design came from one of the set-pattern books that existed.

Further evidence of ideas and/or craftsmen travelling is most acutely shown by a series of misericord carvings that started in Lincoln cathedral in the late fourteenth century. Both St Botolph's, Boston and Chester cathedral have stalls dating to about 1390, and they are remarkably similar in composition and material. There can be little doubt that their misericords were carved by the same man who made Lincoln's. An easy example to take is the story of Yvain and the portcullis gate, which appears on all three buildings, although the story is not exclusive to them since it also appears at Enville, Staffordshire (though, by a different hand). The similarities between Lincoln and Chester are striking, even down to the way the main image on the corbel of the seat is carved and the way the supporting images on each side of the central seat are designed. One can also note the carver's preference for dressing his characters in tightly fitting robes with a series of small buttons.

Likewise, at Worcester and Carlisle cathedrals, there are near-identical misericord carvings made at the end of the fourteenth century. These show a man being swallowed by a dragon-like creature that is sometimes taken to represent Judas in the jaws of Satan. In fact, it is the bigorne, a monster that only ate obedient husbands and so got very fat. Entertainingly, there was also a chichefache, a different dragon that ate only obedient wives and so was painfully thin.[11] On the misericords in question, the wood is different, but almost every other detail is the same, even down to the angle at which each man's foot turns inwards, the pattern on his belt and the placing of the bigornes' feet. This would seem to be evidence of a pattern book travelling; it seems unlikely that someone could have held this image in their minds well enough to have reproduced it so accurately. As is often the case, we do not know which of these carvings was done first. Misericords up and down the land show scenes of domestic violence (always with the woman winning), but the one made at Fairford c.1300 was largely copied by whoever made the one in Tewkesbury Abbey some forty years later. The world of misericords offers numerous examples of ideas travelling; it is like playing a game of medieval Only Connect.

Prior to a rough date of 1200, images were so uniform that it is very difficult to see an individual moving around, although we can see workshops operating and exporting: the late twelfth-century fonts at East Meon, Winchester and Lincoln decidedly come from the same stable at Tournai. Items being exported carried ideas just as effectively as people. Trying to track itinerant artists becomes both feasible and fascinating as time progresses. One of the oldest known misericords dates to 1210 and is in Christchurch Priory, Dorset. It comprises three bunched, closed, acanthus buds, an abbreviated way of carving them that was first seen in stone at Laon cathedral *c*.1204 and which became very popular as a capital decoration. These crocket capitals were not only easier to make than full-blown acanthus leaves, but could be used to symbolize new life becoming eternal life.[12] This image, coupled with the solid bulk of the Christchurch misericords, suggests French journeymen travelling because their style was significantly more solid than the English fashion and their misericord designs lacked supporting images. Likely French workmanship can be seen on the misericords of St Mary of Charity, Faversham, because they are so bulky and also possibly at St Mary's, Wingham, in Kent (where the carvings also lack supporting images), which would make sense since they are so near the south coast. On English misericords there tends to be a central, main image with smaller, supporting illustrations on each side, the subjects of which are not always connected (there are only about half a dozen exceptions to that rule). On the Continent there are either no supporting images, or the design takes up the whole surface.

In other forms it is noteworthy that the minster Chapter Houses of York (1260) and Southwell (1288) lack a central supporting pillar as is the norm elsewhere; someone has travelled with this new technology. It must have been a blessing for meetings of the Chapter who no longer had to peer around a central obstruction. We have the Rheims carvers to thank again in Southwell Chapter House for their fabulous leaves. Their realism is startling – a complete innovation at the time in this country. The Southwell leaves were probably concocted by an Englishman who has most certainly copied English leaves (maple, ivy, oak, and so on) from life, so detailed and natural are they, albeit plainly not to scale. But the idea has equally plainly travelled across the Channel from Rheims, where the capitals are a riot of foliage.

In East Anglia we can see that the same bench-end carver visited both Blythburgh and Earl Soham, whereas the latter shared a doorway carver with Wenhaston. Simon Werman peddled his carved bench panels around Somerset in the early sixteenth century and, most obligingly, occasionally signed them. Sometimes at parish level it is not an individual but a workshop that we see in action. There are many similar fifteenth-century rood screens in East Anglia, several in wonderful condition and showing the amount of money available to church decorators of the period. There was a man who painted charming daisies in the fillets (the strips surrounding the central screen panels), who worked at All Saints, Weston Longville, where he branched out into a tiny, fragile dragon blowing flowers, not fire. However, were one able to put the screens of St Catherine's, Ludham, side by side with that of St Agnes', Cawston, in Norfolk (churches that are only sixteen miles apart, as the crow flies), the similarities would be remarkable. They both have a narrow strip panel behind the heads of some of the saints, alternating background colours and a wavy-line stylized flower motif in the fillets. The Cawston screen, made c.1460, is the younger by a good thirty years, so although it could have been made by the same artist, it is unlikely. While there is nothing to say that a young man should not have produced work of the wonderful quality of Cawston, it seems more realistic that his master would have carried out the work, so we may be seeing the hand of the matured apprentice in the Ludham screen, carrying on the workshop's house style.

Most curious are two very similar images in Norfolk. One is a stone corbel in All Saints, Necton, which dates to about 1410. The other is a misericord in Norwich cathedral made just over a century later. The Necton version of this lively foliate head (also known erroneously as 'the Green Man') is a little fuller in the face and his mouth is closed, otherwise one might wonder if there had been some sort of mould, and perhaps there had because there are two identical versions in stone at Necton, but that would not have helped a wood-carver. When the cathedral staff were asked if they could shed any light, they remarked that it was not unusual for a parish church to copy what was happening in the cathedral, but that is not the case this time; nor are we looking at a hopeful designer practising in one church and moving on to greater things, unless his name was Methuselah. A speculative explanation might be that the carver came from Necton and wanted to have something of his home parish in the headquarter church, but that is unsubstantiated. It remains another example of an idea, a person, a pattern or an item travelling.

What is fascinating to observe as one wanders around our cathedrals and churches is the simple fact that nothing happened in isolation; we may well realize with startled joy that we have recognized someone's work many miles from where we first met him, but that simply would not be possible without his training and sponsorship combined with the generosity, for whatever reason, of the patron. We have seen how Master Masons and their craftsmen had to be alert to changing styles and to incoming influences, and how they often seem to have taken what pleased them most from different sources. Warfare brought Norman (or Romanesque) architecture to Britain and warfare may have been the vehicle that brought Gothic from the Middle East. None of these ideas appear to have travelled as part of a deliberate, grand plan, but at the whim of a bishop, or the ambition of a craftsman. Even so, however the patterns travelled, they are all linked in the great cathedrals we see today and which have been seen by so many millions of our ancestors: they are our most accessible and direct link with our past.

CHAPTER EIGHT

PATRONAGE

THE MATERIALS AND people working on the building site were often paid for out of the Master Mason's own budget, but who found the money for the Master Mason? Without patronage, our villages and cityscapes would look rather different today. Not only was it critical for the funding and functioning of the buildings, but sponsors were more than proud to be seen to be involved. Patrons tended not to have practical knowledge of design techniques, although many took a minute interest.

Again and again, we see evidence that it was not so much a matter of impressing their fellow citizens on this earth, but hoping to find enough favour with God to be admitted to the company of the citizens of Heaven. This could be as an individual operating alone or a group, usually a guild. Some churches are so full of this activity that, were one able to animate the images, the whole building might seem to be full of people waving their arms and shouting to attract the Almighty's attention. It would be rare to find a guide even to the smallest and most humble of our parish churches that does not list an item or a particular part of the building as having been given by someone, or in their memory.

Donor windows

Chartres cathedral has 176 stained-glass windows and, of these, twenty-nine are definitely flagged as being donor windows (two of them modern) with the donors ranging from an individual unnamed priest and Philippe Boarskin (Count of Clermont and Boulogne, son of Philippe II Augustus) to the guilds of furriers, bakers and drapers and others. Other windows were also specifically donated, but the benefactors no longer speak to us. Some capture the imagination more than others and some are useful because they give us evidence of craftsmen at work, usually in a highly stylized way. The Chartres masons donated a window dedicated to St Chéron that shows them sculpting statues of royal figures. The tools of their trade hang above their heads while the stone figures become gradually more detailed as the scene in the window progresses. St Chéron, known in England as St Caraunus, was a deacon from Rome who evangelized the area around Chartres in the fifth century and was killed by vagabonds. The fact that the masons chose to honour a saint from that neighbourhood suggests that they were local labour, as opposed to men imported to work on or oversee a job, or journeymen on their wander years, because otherwise they might have chosen St Clement, who was the most popular of the sixteen saints in charge of stonemasons. At Le Mans, a five-lancet window (an expensive item) was given by the vine growers, and they had themselves depicted in three of the lights. In one of these, they are enjoying baguettes and wine for lunch, just as we might today.

There is a window in the north nave of York Minster noted for its gold and silver bells. This was given by Richard Tunnoc, who died in 1330 and who was a bell-founder in the city and, indeed, one-time mayor. The central parts of the window show the details of his trade: the mould being made, molten metal being poured, the blessing of his work, which was as important a part as anything technical, and the buffing of the finished bell. In many forms of church art, the donors are shown kneeling patiently waiting for redemption, as part of the item they have donated. Sometimes they are not identified, such as the man on a wing of the Mérode Altarpiece dating to the early fifteenth century. Naturally, sometimes the donor is revealed not in image, but in writing. William and Alice Atereth at Cawston, in Norfolk, who sponsored part of St Agnes' remarkable screen in 1490, asked for prayers for those who had had four of the sixteen panels painted, by which we assume themselves. There had already been bequests for this screen from John Barker thirty years earlier, but evidently their joint efforts were not enough because William Howelyn and Robert Osborn bequeathed more funds in 1494 and 1504. Clearly, this was an ongoing community project.

Also commemorated in writing are John Smyth and Joanna Rous in Stradbroke, in Suffolk. A lovely fifteenth-century font was restored in the 1870s under the auspices of the surveyor to the Diocese of Norwich, Richard Makilwaine Phipson who, the guide book says, was reported as having a horror of doing things by halves (*see* plate 14). It depicts an exquisite rendering of the Instruments of the Passion as well as an inscription around the step that reads: *Johannes Smyth et Joanna Rous hunc fontem fieri fecerunt* ('John Smyth and Joanna Rouse caused this font to be made'). Sadly, we know nothing more about what sparked the donors' generosity or how they decided on the images shown.

Where there are images, unfortunately we have no way of knowing if they are exact portraits. In earlier examples, the fashion was for faces to be generic, not portraits. As they moved into the Renaissance period, people became identifiable – or rather, we assume that they are because it would be a good advertisement for the artist if his skill were so good that those who knew the sitters could recognize them. It was also a reason why artists often did small-sized self-portraits: it was an easy and portable set of credentials when they were hoping to persuade a patron that they could paint realistically. That was something that would develop over time.

What we can identify is their belief and desire; every medieval Christian building has someone somewhere saying 'help me', 'notice me', the message not being so much for us (unless it is to ask us to pray for them), but as a last attempt to convince the Almighty that they were worth saving. If we could hear them today, the noise would be deafening.

Grants of land

To start at the beginning, any building needs land on which to stand and off which to feed, in the sense of giving it an income. Some church builders just grabbed and bulldozed (so to speak) what they wanted, as did Bishop Herbert de Losinga at Norwich, but many early charters dealt with setting up religious foundations to the benefit of everybody's souls. One of the earlier ones is that of Pershore Abbey, in Herefordshire, where there was a monastery dating back to the seventh century. In 681, Ethelred, king of Mercia, gave lands to St Oswald, who was then the bishop of Worcester, to set up a religious house at Pershore. Almost three centuries later, this was confirmed by King Edgar (c. 943—75). Likewise, c.956, Oskytel, the archbishop of York, was granted land at Southwell, in Nottinghamshire, by King Eadwig.[1] At Westminster, a man called Ulf, who was the port-sheriff, gave land with his wife in about 1043; port-sheriff of Westminster might seem an unlikely title until we remember how the Thames runs through London and how rivers would have been used to transport goods and people much more than now. The building of the abbey began in the next twelve to eighteen months and, having been granted the land, the work was financed by a tithe on the king's own possessions because he wanted to be buried there. By royal writ, Teinfrith was named as the king's church-wright, and Godwin Gretsyd ('fat purse') is recorded as being both the mason and benefactor.[2] The word 'churchwright' probably means that he specialized in church roofs and timber fittings. In his turn, for his trouble he received a grant of land at Shepperton at some point in the years 1057–66. Godwin and his wife,

Wendelburh, had fat purses indeed because they are also recorded as having given substantially to Hyde Abbey, at Winchester, where their souls were prayed for. They gave land and houses (plural) they owned in Southampton to Westminster Abbey on condition that their son, Ælfwin, had the use of them throughout his life, so they were influential in getting the abbey building work begun. When we walk past many of our most prestigious buildings now we see statues of all sorts of important historical figures, but seldom come away with much impression of people like the Gretsyds, who would have been a critical part of its development. There may well once have been a carved commemoration of their generosity but, sadly, time has eaten it up.

The importance of patronage

Early English kings got actively involved in church building from the beginning and this became very much an English trait; kings saw themselves not only as defenders of the people, but also as defenders of the faith as we would understand it today, although that terminology was not part of their job description then. It went with the territory, almost literally; even kings such as Eadwig, who was not otherwise known for his religious zeal, wanted to be seen to be generous to the Church. He was king for four years, dying in 959, and was known as 'All-Fair' because of his great beauty and because he deserved to be loved. He could not be said to be our greatest king although, to be all-fair to him, he was only about fifteen when his uncle died and he succeeded to the throne. Even though he had bishop, and later saint, Æthelwold as one of his advisors, he still managed to fall out with the equally great saint and then abbot, Dunstan. Later, Eadwig went so far as to send Dunstan into exile, which did not put him high on the good Christian list.

St Dunstan's biographer described Eadwig as a foolish young man 'endowed with little wisdom in government', but he might be forgiven for a little bias. The biographer also tells avidly how, on the night of his coronation feast, Dunstan went to fetch the king, who had absented himself from the festivities. In one of the royal chambers, Dunstan discovered the royal crown, wrought of gold, silver and precious stones, lying neglected on the floor. However, what he also found lying on the floor – though, certainly not being neglected by Eadwig – was not only the king's future wife, but also her mother, which behaviour naturally outraged Dunstan and perhaps set the tone for the rest of Eadwig's reign.[3] Society may have been less prudish than now, but shenanigans with your mother-in-law, designate or otherwise, would have been frowned upon, although it is possible that Dunstan had been just as annoyed by Eadwig's cavalier treatment of the symbols of office, which had been touched by sanctity during the coronation. Eadwig made a large number of land grants to his thanes as a thinly disguised ploy to build up his support-base and gave land to the Church in an equal attempt to curry favour, even if he did undermine these actions by having exiled one of the Church's leaders. In Eadwig's defence, it must be mentioned that he was not hostile to the Church; he supported the reform of the monastery of Westminster and he was a church benefactor, so what more could be asked of him?

In France, land was also given, of course, but it was less likely to be given by the king for the obvious but very simple fact that France was not a unified country until much later than England, arguably not until the reign of Henry IV of France (1589–1610). The man who is acknowledged to be the first king of all England is Athelstan, when he won the battle of Brunanburh in 937. The comparison is not entirely fair since England had distinct boundaries and the set of dukedoms that became the area we now know as France did not. High-powered benefactors abounded, nonetheless. The sister of Pope Callixtus II (Ermentrude de Bar) was living in Autun when her brother visited in 1119 and it is more than likely that the decision to build a new cathedral was made during this visit. The duke of Burgundy gave land and the fact that Ermentrude and her husband were buried in the choir of the church strongly suggests that they gave generously too, but whether in land, money or both is not known. As soon as Nivelon de Quierzy was elected to the bishopric of Soissons in 1176, he gave the cathedral chapter the land on which the south transept was erected; it was very much in the interests of the bishops to create the most beautiful churches, not just for the glory of God, but also for their own status and to increase the diocese's income, which helped their own.

On 8 July 1074 the infantas Doña Urraca and Doña Elvira of Castile gave the church of Gamonal to Don Jimeno, the bishop of Burgos, so that the seat of the diocese could be established in it. This eventually led to King Alfonso VI naming Burgos as the capital of the diocese and ultimately making it the thriving city it is today, as well as a centre for pilgrimage in the Middle Ages. It was unusual for an entire building to be given in that way because it was more likely that a patron would offer a large sum of money, which would act as the trigger for a building made to his or her taste.

Sometimes the land they wanted to build on was already owned by the Church, assuming that they could get permission from the pope. These were Catholic times so his blessing needed to be sought just as the Church of England would have to give authority today for a new-build church within its bailiwick. This was the case at Salisbury in the early thirteenth century when the bishop, Richard le Poore, wanted to build a new cathedral on land the Church owned close to the river. This was to replace the one at Old Sarum that had only been completed in 1092 so, in terms of cathedral-years, it was still quite new. However, one of the decisions of the Council of London, convened by Archbishop Lanfranc in 1075, had been that the hub of the diocese should be relocated to the largest administrative and commercial centre in the bishopric; Old Sarum was not so conveniently placed, nor did it offer room to expand. For the same reasons the headquarters of the diocese of Sussex was moved from Selsey to Chichester. The working parts of Salisbury's new cathedral (quire, transepts and nave) were completed less than four decades after the foundation stone was laid in 1220, which is why it is one of the few cathedrals that looks as though it were made as one piece. Some cathedrals have ended up as a mish-mash of different styles, whereas Salisbury is almost a master-class in Early English Gothic. Since the building was officially finished, it has had remarkably few additions, mainly the tower and spire and a small chantry chapel, but that is quite normal for a building of this status.

Building chantry chapels

It is through chantry chapels that we so often see the main patrons and, in these cases, it is usually the patron's name that survives when the name of the artist or sculptor has been lost to us. A high-status church might have numerous chantries since not all of them involved specific building or alterations to the fabric of the church; they could just as easily be placed within what already existed, because their purpose was to be a focal point where Masses where sung or chanted to help the patron's soul through Purgatory. It would be normal to request or require numerous Masses. It seems that no one thought that just one Mass would do; also, a continued interest by many people was considered to be an important part of the procedure. Some Masses were to be sung or said in perpetuity, while other people laid down a specific number in exchange for a donation that went to help the longevity of the building. At Christchurch Priory, in Dorset, a constable of the castle, Sir Thomas West, died in 1406 and gave the priory £18 18s 4d in exchange for 4,500 Masses to be said for his soul.[4] We do not know if he had a particularly troubled conscience, but we can work out his thinking: £18 18s 4d equals 4,540 pence so he gave 1d per Mass, plus 40d. We have no idea why he hit upon 4,500 in the first place unless that was the maximum he could afford, but we can assume that he was taking inspiration from the Bible for the extra 40d. Several notable characters spent forty days in the wilderness ranging from Noah at sea, Moses (twice), Elijah and, of course, Christ himself, so we can imagine that Sir Thomas saw himself ending up in the wilderness of Purgatory and hoping that this would help to get him through it. He was also following in his mother's footsteps since Lady Alice West, who had died only eleven years earlier, had left a slightly smaller sum for 4,400 Masses to be said for her soul. If she followed the same principle, she must have left £18 6s 8d.

We can also see that patrons of chantries wanted memorials to be in several places where they had themselves lived or worked, not necessarily just the last place or the most significant one. Edmund Audley is a case in point. He was probably eighty-five when he died in 1524 having held numerous ecclesiastical offices, including those of bishop of Rochester (1480), bishop of Hereford (1492) and, finally, bishop of Salisbury in 1502, where he died twenty-two years later. Even so, when he made his will on 11 June 1523 he left provision for a chantry at Hereford cathedral in the form of a beautiful screen behind which was a chapel where Masses would be said for his soul.[5] He also had a chantry chapel at Salisbury, as one might expect since he was buried there, having already given £400 to Lincoln College, Oxford (from where he had first graduated in 1467) to make sure that they observed his obit (anniversary of his death) there. He evidently hoped that he could buy his way into Heaven. Individuals presumably thought that they would please God by adorning the church, although chantries can have a more complex history because they might have duties of care or teaching attached to them.

Bishop Audley was a fervent heretic hunter, especially in the last two decades of his life. While he was at Salisbury there were seventy prosecutions of people who were Lollards in some form or another and some of these ended in burnings. Lollardy was another spin-off from the plague years in that people did not understand why and how the clergy had not been able to foresee the disaster and avert it, and so some wanted to read the Bible for themselves in a language they could understand – English, not Latin. To us this sounds a perfectly reasonable and rational wish, but to the Church hierarchy at the time it was heresy and needed to be stopped. The clergy were aghast at the thought of the untrained minds of the grubby rabble picking over the Word, which was just as much of a relic as a physical piece of Christ. This was not entirely to do with power, but also a genuine concern that proper respect should be paid, not least because of the equally genuine fear of divine retribution if the Word was bandied about or misinterpreted. This was, after all, how many thought that plague had come among them in the first place.

Audley saw himself as a protector of the sacred texts, becoming an enthusiastic 'winkler-out' of those who abused them, and there is no doubt that some of his methods would be repugnant to us today. He once had an eleven-year-old boy hauled up in front of him because the child had been heard discussing Lollard doctrines with other children. The child is reported to have:

said publicly when there were others present playing with him that the body of Christ was not in the sacrament of the altar – and also that images of saints were not to be adored because they were made of stones.[6]

When asked where he had learned these ideas from, the boy was obliged to admit in the end that he had heard these from his father. (One is reminded of W. F. Yeames' famous painting in Liverpool's Walker Art Gallery, 'When Did You Last See Your Father?') Audley will have gone to meet his Maker with a clear conscience because he felt that he had done his Christian duty thoroughly.

Chantries are associated with many of the great families of the day, and it was not uncommon for them to become a family vault, so patronage further became linked with death. At Bristol cathedral there is not only a double chantry, known as the Berkeley Chapel, which has two altars and was founded after the death of one of the Ladies Berkeley in 1337, but other members of the family are commemorated elsewhere in the cathedral. Maurice, the ninth Lord Berkeley, and his mother are buried in the elder lady chapel on the opposite side of the cathedral ('elder' because it was built some fifty years before the eastern lady chapel) and the whole family are very much in evidence throughout the building. Tewkesbury Abbey has three chantries asking for prayers for the Beauchamp and Warwick families. Isabella Despenser had two husbands, both of whom were disconcertingly called Richard Beauchamp. They were cousins (albeit distant ones) of the same generation. One was the Earl of Worcester and the other that of Warwick. In 1430, Isabella built an exquisite chantry as a memorial to both her husbands and to herself. Looking out from its lower ceiling, it has a serenely lovely woman's head, which is presumed to be the lady herself.

As far as patrons' portraits in chantry chapels go, very few can cap the so-called Kneeling Knight at Tewkesbury above the chantry of the Holy Trinity. This is another of the Despenser family, Edward, who died in 1375, although the chantry was not built for another fifteen years. He is shown in full armour and in an attitude of prayer with his eyes open. This is usually taken to be the sign of a hypocrite in medieval iconography, but, in this case it is almost certainly meant to demonstrate his clear, steady, grey-eyed gaze. He is the epitome of a medieval gentleman and knight, with no blemish to his perfect skin, classical facial bones and a manly, athletic figure. He was one of the early members of the Order of the Garter, which Edward III began in 1348 (in 1361 Edward le Despenser was the thirty-eighth member to be appointed). The Order was commensurate with everything chivalrous and it is thought that Edward III was romantically looking back to the Arthurian Knights of the Round Table as his model; indeed, recent excavations at Windsor castle have uncovered what could be a round building where they held their banquets. The Order may have taken its symbol from an incident at a ball held in Calais to celebrate the capture of that town when the Countess of Salisbury's garter fell off and, to spare her blushes, the king is said to have picked it up, tied it on his own arm or leg crying: *'Honi soit qui mal y pense* ('Shame be to him who thinks evil of it'). The story is probably a fiction, but the Order remains the senior Order of Knighthood in Britain, and Edward le Despenser seems to have been a model of chivalry within it.

The medieval chronicler, Jean Froissart, came to England the same year that Despenser was admitted to the Order and was lucky enough to secure him as a patron. This may have influenced his glowing reports revealing Despenser to have personified everything desirable in a medieval knight. Nonetheless, Froissart tells us when writing about what we call the Hundred Years' War:

> The Constable of the whole army was Edward, Lord Despenser, one of the great barons of England, a spirited, courteous and gallant knight and a fine leader of men, who had been appointed to the post by the King himself.[7]

Despenser was much admired by the ladies, and it was written that 'the most noble said that no feast was perfect if Sir Despenser was not present'. He died young, aged thirty-nine, succumbing to a disease that wiped out many in the English army at Bordeaux in the winter of 1375. It is not hard to imagine that he was 'deeply mourned by all his friends' or that he had 'a noble heart' and was 'a gallant knight, open-handed and chivalrous'. The image of the knight at Tewkesbury certainly fits the mould, although we cannot know how accurate a portrait it is, not least because it was made so long after his death.

Many had their tombs constructed long before they died. That way they could be quite sure that they would be commemorated in the way they wanted to be. More importantly, they had a focal point that they could use to help them dwell on the arrival of the Grim Reaper and how it would ultimately affect them. This was considered to be spiritually healthy, whereas nowadays we might find it morbid. Henry Chichele, the sixty-second archbishop of Canterbury (1414–43) commissioned his tomb some eighteen years before his death; although he could not have known when he would die, such forward planning was not uncommon. Today many people buy burial plots well in advance and plan and pay for their funerals, but the medieval trend was to go a step further and remind oneself that, as Shakespeare's Guiderius puts it:

> Golden lads and girls all must,
> As chimney sweepers, come to dust.[8]

Death was never far from the medieval mind, much emphasized by the onset of plague, and we can see the evidence for ourselves in our cathedrals and churches. What is special about Chichele's tomb is that it is a double-decker: on the top he is wearing his fine robes of office, but down below he is shown as the wretched cadaver he would become, shrinking back to earth. Chichele had had his tomb created like this so that he would never forget that one day – any day – he could be called in front of his Maker and asked to explain himself. He was not alone in his thinking: at Tewkesbury, the Wakeman Memorial is, in fact, a cenotaph but, in terms of image, it makes the same point, as do the cadaver tombs at Lincoln, Waterford, Winchester, Tideswell and Breedon-on-the-Hill, among others. In a way, Isabella Despenser did the same thing in that the memorial for herself and her two husbands was made before she died. Indeed, if the image of the serene and lovely-looking head is of her, she remains as lovely to this day as she was in life.

The memorial-makers

These tombs, chantries and memorials were, obviously, all designed and made by someone, and very often someone of considerable skill. By no means are these apprentice pieces merely because they are smaller in scale, although inevitably some of these were the stepping stones enabling a promising craftsman to be elevated to the rank of Master Mason. The names of many memorial-makers were not recorded because it was the person commemorated who was important, but several Master Masons specialized in tomb-making. In 1297, Michael of Canterbury, the same man who stood surety for Michael de Hoo and others for their botched work at Eltham (see Chapter 5), made the tomb for Aveline and Edmund 'Crouchback', Earl of Lancaster, which survives intact in Westminster Abbey, as well as the tomb for Bishop William of Louth in Ely cathedral. He designed many other memorials as well, but perhaps nothing as prestigious as the Cheapside Eleanor Cross at a cost of £226 13s 4d from 1291–4. This was the second most important of the twelve crosses commissioned by Edward I when his wife, Eleanor of Castile, died in 1290, at Harby, near Lincoln. A cross was erected at each of the places where her funeral procession stopped on its way to her final resting place at Westminster. There, she was laid to rest in a tomb designed by Richard Crundale, who had worked on the Charing cross with Ralph of Chichester (a marbler who also worked on the St Albans and Woburn crosses). In fact, Eleanor did not arrive in London intact, as was so often the case in those days: her heart was buried at Blackfriars, in London, whereas her viscera remained in a tomb at Lincoln, designed by Nicholas Dyminge. Building these tombs were extremely high-profile jobs to get, catapult-ing the craftsmen involved to an altogether higher level of patronage. Of these crosses, only those at Geddington and Hardingstone survive completely, whereas the one at Waltham Cross has been much restored and the one outside Charing Cross station is a Victorian replacement. Sadly, the charming story that Charing Cross had its name from the grieving king referring to the final cross being for his '*Chère Reine*' is untrue: the hamlet had been called Charing long before the cross was constructed.

Royal patronage

Henry III was the great building king. As Lethaby says, London was practically transformed during his reign, with works ranging from the spire of the then St Paul's, the gates of the City and the Tower, projects at Westminster and Kennington as well as at Winchester, Windsor, Clarendon, Gloucester, Guildford, Woodstock, and many others. He was not the only royal patron, but he was the most prolific. Unfortunately, royal patronage did not always work, as Jocelin of Brakelond recalled:

> King John, ignoring all his other commitments, came to St Edmund's immediately after his coronation, impelled by a vow and out of devotion. We naturally thought that he would make a sizeable donation, but he gave only a silk cloth which his servants had on loan from our sacrist – and still they have not paid for it. Although he had accepted St Edmund's most generous hospitality, when he left he contributed nothing at all honourable or beneficial to the saint, except the thirteen shillings which he gave at Mass on the day he left us.[9]

There is a real and not unreasonable note of grievance there. It was expected that a wealthy or royal visitor would make a substantial gift. This might not offset the cost of his visit because it could not always be sold, but at least it might attract visitors and pilgrims who would donate, or just make the monks feel that their efforts were appreciated. After all, when Cnut and Queen Emma visited Sherborne Abbey, in Dorset, as far back as 1010, making a pilgrimage to the shrine of Bishop Wulfsin after his translation, Emma had given twenty pounds of silver for the repair of the roof above the bishop's tomb because it was 'disrupted with age and gaping with cracks'.[10]

One would have to have been a real saint to have given hospitality and not to have counted the cost in the wake of a royal visit. Ideally, a royal guest would donate a practical gift of money for a new chapel or administrative building, but even to have his ongoing interest and involvement could make an enormous difference to the building because wherever the king went, sycophantic others were sure to follow. King John does not seem to have grasped this vital point.

Royal patronage also failed under Edward I at Vale Royal Abbey. In 1277 he wanted to improve the Cistercian abbey there, stipulating that it was to be bigger and better in every way than his grandfather's (King John's) now ruined abbey at Beaulieu and ditto his uncle's (Earl of Cornwall's) at Hailes, which is now also a ruin. It was to house one hundred monks.[11] Perhaps family rivalry was not the soundest form of motivation and, in any case, Edward got distracted by the need to build castles for defence, which ate into the building budget. Vale Royal Abbey progressed in a sporadic way for a while but, if the patron's heart is not in it, then no one else's will be. This must have been dispiriting for the local community and exasperating for the Master Mason, who may well have given up the chance to do other prestigious commissions for the opportunity of having a royal commission on his CV. After some thirteen years, the king's agent at Vale Royal was told that Edward had 'ceased to concern himself with the works of that church, and henceforth will have nothing to do with them', leaving the whole works high and dry. Over the years Edward II and the Black Prince dabbled at Vale Royal, but their efforts too were less than wholehearted. In 1336, when the building should have been more or less complete with a happy, thriving community of monks living in leak-free, brand-new accommodation, with no excuses not to worship without distraction, the then abbot wrote despairingly:

> We have a very large church commenced by the King of England at our first foundation, but by no means finished. For at the beginning he built the stone walls, but the vaults remain to be erected together with the roof and the glass and other ornaments. Moreover, the cloister, chapter-house, dormitory, rectory and other monastic offices still remain to be built in proportion to the church; and for the accomplishment of this the revenues of our house are insufficient.[12]

The inevitable happened on 19 October 1360 when there was a colossal storm and the nave collapsed, probably because the wind could get into it. The destructive force of the wind is often overlooked until far too late and when battening down the hatches is about as effective as mopping up the sea with a piece of cotton wool. In Chapter 2, we saw how destructive high winds could be at Soissons. People occasionally marvel that such great abbeys and cathedrals can end up as rubble-strewn ruins and disbelieve the amount of money it takes to maintain these buildings every year. Just to keep the wind at bay is a remarkable achievement in itself. In September 1983, I once witnessed a similar incident in a Hong

Kong tower block during Typhoon Ellen, when a window had not been made safe
on the thirtieth floor: the wind soon blew in the glass and subsequently ripped
out all other glass; those of us watching from a more secure building opposite saw
not just in-trays and files, but whole desks, chairs and even a filing cabinet flying
through the air like toys tossed from a dolls' house.

Other important patrons

From this we can see that it might sometimes have been better to have a non-
royal patron who had fewer calls on his purse. In cathedrals, especially, many
people were giving at some level or another. Some gave whole chapels, while
others donated practical furnishing, such as Abbot Elyot, who gave the choir
stalls and misericords to Bristol cathedral in 1520. Many were patrons in the
sense of either getting a venture off the ground or by driving it through to as close
to completion as any cathedral ever is. It was an ongoing task and it is sometimes
surprising to see the sort of people who were giving, as the following extract from
Jocelin of Brakelond's account shows. A lot of building was going on at Abbot
Samson's instigation to improve the amenities at Bury St Edmund's.

> The chapels of St Andrew, St Katherine and St Faith were newly roofed with
> lead. Also many improvements were made inside the church and out, if you
> don't believe this, open your eyes and look ... At this time, too, our almonry
> was rebuilt in stone - formerly it had been ramshackle and of wood. Towards
> this our brother Walter the physician, then almoner, gave a large donation of
> money that he had made from his medical practice.[13]

In addition, dismayed that the silver frontal of the high altar and many other
precious ornaments had been disposed of to pay for the recovery of Mildenhall
and for the ransom of Richard I, the abbot did not wish to refashion the frontal
and other similar panels in case they should suffer a similar fate.[14] Instead, he
focused all his efforts on creating a precious canopy to set above the shrine of
the martyr Edmund. He wanted this work of art on display in a position where
no man would dare to lay a hand on it and from where it could not be removed
easily. One might not have expected a physician to have so much money that he
could make a substantial donation, especially if the words 'our brother Walter'
denote a monk, as opposed to a social nicety. Appointments in abbeys such as
physician were filled by monks who had particular expertise or training and as
monks swore an oath of poverty (amongst other things), he should not have been
wealthy. This man was also the almoner – the distributer of alms to the poor

– which further indicates that he was a monk, so a mystery remains as to the origins of his wealth.

The ransom for Richard I created real hardships for an already hard-pressed English people. In 1192, Richard had been returning from the Crusades when bad weather had driven his ship ashore near Venice. Captured and imprisoned by Duke Leopold of Austria, Richard was then handed to the Holy Roman Emperor Henry VI, who demanded the sum of 150,000 marks for his release. The king's ransom was so much money that the expression moved into the English language. It was certainly a substantial blow for the inhabitants of this island, who could be forgiven if they felt less affection for the king than before. If a mark was 13s 4d, 150,000 marks would be £100,000, an amount apparently weighing three tons in silver. Even if the mark was reckoned at 6s 8d, as it sometimes was, it was still a considerable fortune. To find this sort of money a tax of two shillings per hide was introduced, both clerics and laymen were taxed at a quarter of their rents and goods, England was surrendered to the emperor as a fief, and anything that could not be taxed was melted down for bullion.[15] There are reports of religious buildings being stripped of their valuables in order to redeem this absentee king, which is what was worrying Abbot Samson and makes it all the more remarkable that all works did not stop altogether, although many did. Church leaders often managed to find a way to protect their beloved buildings as this account by Jocelin of Brakelond shows:

> Nevertheless, whether St Edmund's shrine should be partly stripped for the king's ransom was argued before the Barons of the Exchequer, and the abbot stood up and answered the point in this way: 'Take it for a certainty, that this shall never be authorised by me, nor is there any man who would get me to agree to it. But I will open the doors of the church – let anyone enter who will, let anyone come near who dare.' Each judge replied with an oath, 'I shall not go', 'Nor I. St Edmund vents his rage on the distant and the absent: much greater will his fury be on those close at hand who seek to rob him of his clothing.' Because of what had been said, the shrine was not despoiled, nor was a loan raised on it. So putting other matters aside, the abbot decided very wisely and sensibly to construct a canopy for the shrine. And now sheets of gold and silver resound between the hammer and the anvil.[16]

This is interesting as a separate study in itself because it shows the power of the saints as perceived by the thinking people of the day. As this ruse seems to have worked so easily, one wonders why more religious leaders did not deploy it to save their treasure; it also tells us about the character, leadership and persuasiveness of Abbot Samson.

The payment of tithes

The payment of tithes might be thought to be a good way of funding a building. This was a system that goes back into the mists of time, the first reference to it being at a synod of 786 and later in 900. The idea was based on several Old and New Testament references, not least St Paul's injunction to us all that God loves a cheerful giver. Originally, it was a way of paying for people in the sense that a quarter of the offerings of the laity went to the priest, with the remaining three-quarters going to the upkeep and fabric of the church, to the relief of the poor and to the bishop. At some point it must have become clear that random and general offerings needed to be defined as a specific amount and so compulsory payments of a tenth part of all the produce of lands came in. It is worth noticing that the fabric of the building was seen to be something to be supported, but this seems more likely to be in the sense of maintenance rather than new building. At first the landowner could pay the tithe to whichever clergy he pleased, or even to the bishop, who was expected to redistribute it. The system was open to failure whether from personal greed or just incompetence in collecting the dues, but it was attractive, not least to the local landowner, who could use the monies raised to build a church and, therefore, to bring himself to the attention of the Almighty.

In time this allocation of tithes became the law and it evolved into a way of funding clergy salaries. One of the quirks of tithes was that they could be great or small depending on whether they were collected from the main cereal crops or from animals. Crops were more attractive because a tenth of cereal was easier to quantify than a tenth of an animal. If the incumbent was entitled to the whole tithes of a parish, he was termed a rector. He usually then endowed a priest who lived among the people of the parish and was their personal spiritual adviser. This man usually took a portion of the glebe together with the small tithes (animal produce) that were more troublesome to collect; he was known as the vicar. Not every part of England was obliged to pay tithes, but it gradually evolved so that a set amount of money was paid annually in lieu of a tithe. Ultimately, this was fixed by legislation, the amount being dependent on varying local prices of corn, and was not abolished as a system until the 1930s. Vestiges of the tithe system can be found to this day in the parish share or quota paid annually to the local diocese. Plainly, it is now only those who attend church who pay, each parish's share being based on the number on the Electoral Roll and moderated by whether or not they have a full-time priest, a shared one or even none at all. The amounts are calculated differently in each diocese, but today the money essentially contributes to the costs of clergy stipends and, more recently, to their pension contributions. That still leaves individual congregations

endlessly fund-raising and half-hoping for benefactors who will help with maintenance, running costs and meeting the seemingly endless bills associated with
running a community asset in the modern world.

The clergy as patrons

In Chapter 2 we saw how bishops would give out of their own pockets in some
cases, whether or not they really wanted to. Perhaps the greatest benefactors
among such donors were those who did not just pay in money (although that
was always welcome), but in motivation to build, sometimes bringing fresh
ideas to the design. Durham offers several examples. If you look at this delightful cathedral from the north side on Palace Green, the central section with
a rounded, Romanesque or Norman arch design is the oldest extant part. It
was the brainchild of William Saint-Calais (also known as Carilef), who was
bishop from 1080 until his death in 1096. The whole of this building stands as
memorial to Saints Cuthbert, Bede and Oswald; the parts of it are individual
expressions of piety and bring equal joy.

William Carilef/Saint-Calais had begun his religious life as a Benedictine
monk at the community of Saint-Calais, which is roughly twenty-five miles east
of Le Mans, in France. In 1080, he had been appointed bishop of Durham by
William the Conqueror and one of his early decisions was to replace what is
known as secular (non-monastic) clergy with monks. He discussed this with the
king and Archbishop Lanfranc before going to Rome to get permission from
Pope Gregory VII to proceed. This was granted in 1083, so he expelled the clergy,
some of whom were married (perfectly legally then since the celibacy of clergy
was not instituted until the twelfth century). Some of the monks he brought in
as staff came from Jarrow, which had recently been reinvigorated as a monastery. Jarrow was where Bede had been buried in 735, but in 1022 his bones had
been stolen and taken to Durham, thus establishing a link with Jarrow. Bede's
bones, along with Oswald's head, assorted other bones and a number of exquisite
items had all been put into St Cuthbert's coffin, which by now was lodged in the
first cathedral on the site. William Saint-Calais had this building demolished
to make way for the beginnings of the current one, which was then built in the
Romanesque style (also known as the Norman style), which was finding favour
in this country. This all happened very quickly and we have a record of the new
bishop leading his Chapter in a service of dedication on the site on 29 July 1093,
with the first stones being laid a fortnight later (11 August). The motivation for
the new cathedral was not, of course, for the aggrandisement of Saint-Calais, but
for the safe and suitable storage of St Cuthbert, the most powerful and best-loved

saint of the north of England, who had died on one of the Farne Islands in 687.

Saint-Calais had an eventful final decade of his life. After the Conqueror died in 1087 there was the inevitable shake-down into the reign of his son, William Rufus (William II), and this included a rebellion against Rufus in 1088. Rufus was convinced that Saint-Calais was involved and so, first, put him on trial in Salisbury, then exiled him back to Le Mans. Saint-Calais was back in favour and again working at Durham in 1091, but it may be that his time back in his native land gave him a chance to reflect, to look at the latest cathedral designs in Normandy and Maine, and to give him ideas that he could bring back to England. The new cathedral at Durham was to be larger and more noble than the one that had preceded it, and it was paid for out of his own pocket. He died before his project could be realized, although the monks made sure that building works were carried on as much as they could. The next man on the scene was Ranulf Flambard, bishop from 1099–1128.

Flambard was said to be impatient of leisure, among other less flattering descriptions. One of his claims to fame is that he was probably the first man to escape from the recently built Tower of London in 1101. At this distance it is hard to tell, but he seems to have been a man who inspired either devotion or odium, which may not be unnatural for one of William Rufus' chief administrators. He was certainly a very able man, highly intelligent, quick-witted, and he liked to build. Before he arrived in Durham he had caused the first stone bridge in London to be constructed (a forerunner of the modern London Bridge), a wall around the White Tower and a new hall at Westminster. Seeing what William Saint-Calais had begun, he almost immediately ordered that the land between the cathedral and Durham's Norman castle be cleared to make a level space, creating the area known today as Palace Green.

It was under Flambard's jurisdiction that St Cuthbert was translated to a new shrine within the new cathedral, one that would do justice to his status. Symeon of Durham's account, written in the twelfth century, is not without its lighter moments as it seems that Flambard began a sermon outside the east end of the cathedral where:

> he kept preaching on, touching on many points not at all appropriate to the solemnity and fairly wearing out the patience of many of his hearers by the prolixity of his discourse.[17]

It seems that even the long-dead Cuthbert got bored because there was a sudden torrent of rain, causing all to scatter and the coffin to be rushed indoors where it was found not to have a drop of rain on it.[18] It was at this same time that Cuthbert's coffin was opened because of rumours and doubts circulating about whether or not his body could possibly be intact after some 400 years. Two layers of coffins were found around him, as well as other skeleton remains and priceless items such as the pectoral cross and diminutive Gospel of St John, which is now in the British Library. More importantly for the witnesses at the time, when the lower coffin lid was lifted they:

> smelt an odour of the greatest fragrancy, and behold, they found the venerable body of the blessed father laying on its right side in a perfect state and from the flexibility of its joints, representing a person asleep rather than dead.[19]

After Flambard, there came and went three bishops, under whose jurisdiction we assume that the building progressed as well as a building does when it is has lost a major patron, so the next noteworthy patron in terms of construction was Hugh le Puiset (also known as Pudsey), a great-grandson of the Conqueror. He did a lot of work on the castle, built Elvet Bridge and funded numerous other projects in the north east, notably at Norham. He wanted Durham cathedral to have a Galilee, which is, in effect, an imposing porch at the west end. It is not found very often in Britain, although there are examples at Lincoln and Ely. The idea of a large porch evolved from the original narthex, but it represents the journey that Christ made from Galilee to Jerusalem, hence its name. It is multi-functional, acting as a space where processions form up and meetings take place and, occasionally, a consistory court. It encloses the west doorway, in Durham's case making it impossible to enter the cathedral other than from the north side or through the cloisters. Its dramatic setting above the River Wear makes it seem as though the Galilee is built over a precipice (it is not).

Puiset had wanted to build a lady chapel on the east end of the cathedral, but as the walls rose higher, more and more cracks kept appearing. The bedrock at that end of the church is uneven, hence the problems with cracks, but it is also very close to where the new shrine of St Cuthbert lay. This led to the notion that Cuthbert was a misogynist who would not tolerate a woman near him, even if that woman was the Mother of God, so it was he who was causing the problems. That further led to a black marble line being inserted into the floor at the west end of the cathedral (just in front of the font today). No woman was allowed to cross this line for fear of offending St Cuthbert, a rule that still seems to have been in play in the fourteenth century, if not later. There is no basis in history at all for poor St Cuthbert to have been maligned in this way! There is a thought that the Galilee might also have been built as an unusually placed lady chapel since they are generally found at the east end. The Durham Galilee is a remark-able design having five aisles and lobed arches that are reminiscent of ones found in Cordoba or Seville – further evidence of ideas travelling. It now houses Bede's tomb, courtesy of Puiset, who separated out the bones in Cuthbert's coffin and brought together those that he thought most likely to have belonged to Bede. Bede was given a garish-sounding shrine of gold and silver, but in 1831 was re-housed in a plain black tomb.

We know a little about the Master Masons who worked for le Puiset, the chief of whom was Richard de Wolveston, an engineer who was in Durham from 1170–82. He was held to be a very high-grade craftsman much admired at the time both for his work and character, so much so that one of the monks gave him a fragment of the cloth in which St Cuthbert's body had been wrapped. Wolveston put this away carefully with some 'painted letters which he possessed of very great beauty'. There are no further details of these, but they may have been illuminated letters, perhaps used as templates. He accidentally left this wallet in an inn at Berwick-on-Tweed after a visit to Norham. Precisely what happened is not clear, but as the outcome is described as a miracle, it is assumed that he got the wallet and contents back. Then there was Lambert the Marbler, who is recorded in 1183 working on marble columns for the Galilee. As part of his contract he was able to rent thirty acres in Stanhope at 4s a year while in the service of the bishop.

Richard le Poore arrived in 1229, the last of the medieval builder-bishops for Durham. An observant visitor at Durham said he was reminded in some way of Salisbury when he reached the Chapel of Nine Altars at the east end. He was on the right stylistic track because Richard le Poore had been posted from Salisbury with the best part of a decade's experience of building the new cath-edral there under his belt. He was responsible for expanding Durham's cathedral, and the chapel spreading into eastern transepts was modelled on the now ruined Fountains Abbey, which he had visited. The Nine Altars are notably shallow in depth, presumably because le Poore had learned from previous failed attempts to build at the east end. Sadly, le Poore died before seeing the Nine Altars come to fruition.

A few years later, Nicholas Farnham was consecrated bishop in 1241. It is worth noting that the 'Architect of the new fabric of Durham' – indeed, the one who probably worked on the Nine Altars – was one Richard de Farnham, appointed in 1242, so perhaps they were related. Bishop le Poore was one of numerous churchmen who took a close interest in the design and improvement of their workplace; it is hard to think of another profession that has done so to the same extent and achieved such spectacular results. Durham has been taken as a small case study, but equally one could cite Ely, where bishop Hothman spent £2,035 on the choir where he wanted to be buried, or bishop Northwold giving over £5,000 in the 1240s. This covered almost three-quarters of the total cost needed to rebuild the east end. With the senior churchmen, it is not always possible to spot the dividing line between building for the glory of God or for themselves, or for the sheer joy of being able to create such eye- and mind-boggling structures.

It was standard practice for an abbot or someone senior within the hierarchy to give. It may have been in the sense of a substantial building, or it may have been in the sense of smaller things; perhaps they enhanced one of the facilities of the establishment, as did Abbot Simon of St Albans. The Chronicles of St Alban are especially helpful in listing the works of several abbots and the passage below describes Simon's generosity:

> Abbot Simon [d.1183] of immortal memory honourably supported two or three scribes in his residence; he made careful arrangements to have a priceless supply of excellent books and he stored them in a special cupboard. He also restored the scriptorium, at that time largely dispersed and neglected, and introduced in it certain praiseworthy practices: also, he increased its revenue so that in all succeeding times the incumbent abbot would continue to have a special scribe. Apart from the very valuable books, which would take too long to enumerate, he presented silver basins and many other vessels and ornaments to God and the Church in bestowing them on the Blessed Martyr.[20]

As today, these buildings depended on personal generosity, but in the Middle Ages many of those sponsoring cathedrals, whether in money or energy, left their mark, not in the form of masons' marks or anything to do with the technicalities of building, but sometimes as a rebus on a benefactor's name. Medieval man loved his puzzles and riddles and visual puns as much as anything; so many remain to this day for us to work out. Even if we fail to delight in them in the same way, as we stroll around a cathedral they are useful 'memory joggers' as to who the main patrons were. Some of the more obvious ones are those of Abbot William Burton, who donated the parapet that finishes the screen at Bristol cathedral (c. 1530). Apart from his initials, it shows a burr (thistle) and a ton/tun (barrel). Canon John Swinfield of Hereford Cathedral has several swine wearing his colours in a field and Bishop James Goldwell's rebus at Norwich hardly needs to be described, although he did cause ninety-four of the bosses in the presbytery, which he re-vaulted in 1463, to contain gold wells (just in case we missed his generosity). Sometimes a double memorial with patrons honouring patrons can be seen, for example, at Beauvais cathedral there is a window given in 1522 by Françoise d'Halluin, who is shown as an elderly lady looking a tad worried the nearer she gets to meeting her Maker. Also in the picture is St Francis, her name-sake patron, so we can see her inspiration and it helps us to remember her name for our prayers. This was a very common device, with some parish churches being dedicated to the saint after whom the main benefactor had been named, which might account for the world-wide shortage of churches named for St Zoticus, although at least nine saints suffered that name. In Bristol cathedral there is a screen given by a Thomas White in 1542, which he acquired from a Carmelite church after the Dissolution of the monasteries. Post-Reformation, the days of signs, symbols and possibly even frivolity started to decline, so Mr White settled for his initials along with his merchant's mark, which is equally interesting, but not as much fun.

CHAPTER NINE

HOW DID THEY DO IT?

The building process

THIS BOOK IS not a simple 'how-to' guide for building a medieval cathedral. We have already seen that very few masons' manuals, if there are any worthy of the name, have survived. This is likely to be because very few such manuals were written in the first place. We have also seen that guilds encouraged confidentiality, especially the keeping of trade secrets, and that Masters in Regensburg, at least, had to swear an oath of secrecy. Indeed, Mathes Roriczer, whose work we will now consider, was accused of breaking this vow simply because he explained in writing how to build certain things. However, as Lon Shelby has suggested, Roriczer was operating at a time when Master Masons were becoming more like modern architects; primarily designers who were increasingly (but not entirely) leaving the execution of the design to craftsmen.[1] We have a lot of contracts and other evidence that tells us who wanted a building to be built and, occasionally, why, how much they were prepared to pay, and so on, but not often anything that tells us how the Master Masons set about the business in terms of physically putting the thing together. You could simply say, as Isidore of Seville did in his *Etymologies* of the seventh century, that:

> buildings have three parts: the plan, the construction and the aesthetic considerations. The plan is the description of the site or ground and of the foundation.[2]

No one is likely to dispute that.

Occasionally, a contract sheds some light on the building process, such as the one for the tower of the church at Arlingham, in Gloucestershire. The contract was made on 25 November 1372 between nineteen named parishioners, who included the vicar, Roger, and all parishioners of Erlyngham, and the Master Mason Nicholas Wyshonger of Gloucester. The parishioners contracted with Wyshonger to complete the bell tower of their church within the next three years, building 12ft (3m) each year. He was to: set corbels inside the walls to support the various floors; make a door on the east side, leading into the roof; create a handsome window on the west between the first and second floor levels; and make a small window in each wall of the uppermost storey. At each floor level there was to be stone tabling, and, on the top, battlements with gutters. He was also to make a spiral staircase with doors at top and bottom. The parishioners were to provide all materials, except tools, and were to bring them to a suitable place within forty feet of the tower.[3]

While this contract does not tell us how the workers set about the job in the sense of measuring up and mixing mortar, we do get an impression of order and careful monitoring. One wonders if whoever led the parishioners (not necessarily the vicar) had had a bad experience before with Master Masons. It is precise and has an air of being the result of much discussion. The words 'handsome window' mean nothing to us, but they smack of additional expense. Perhaps Nicholas Wyshonger had had a bad experience with a previous client; we must assume that it is he who stipulated that the materials were to be brought within a set distance. Had he spent a previous job grinding his teeth while going forwards and backwards to a distant heap of stones? The results can still be seen today (*see* plate 15).

This poem was written around 1245 when Westminster Abbey was being rebuilt by Henry III, which gives us some clues as to the overall actions:

Now he laid the foundations of his church
With large, square blocks of dark stone;
Its foundations are broad and deep,
The front towards the east he makes round,
The stones are very strong and hard.
In the centre he raises a tower,
And two at the west front,
And fine and large bells he hangs there.
The pillars and the cornices
Are rich without and within
From the bases to the capitals
The work rises grand and royal.
Sculptured are the stones,
And storied the windows;

All are made with the skill
Of good and loyal workmanship;
And when he has finished the work
He covers the church well with lead.
He makes then a cloister and a chapter house
With a front vaulted and round towards the east,
Where his ordained ministers
May hold their secret chapter.
Frater and dorter
And the office round about it.[4]

That gives a sound overall plan, but it would not help a group of masons and carvers standing on a prepared piece of land, wondering precisely how and where to start. There is a glimpse of how the land was prepared, not just in the sense of clearing luckless town-dwellers off it in the first place, as at Durham. Alexander Neckham lived from c.1157–1217. He was a well-educated man who had taught in Paris as well as St Albans and who ultimately became the abbot of Cirencester. He did write a manual, sadly not on building, but on the nature of things (*De naturis rerum*), which dealt with scientific knowledge of his time. Neckham has much to say about mankind's insatiable vanity and affectation which, he says:

> Is shown in part by the expenditure dedicated to pleasure which an empty boastful pride consumes and squanders in the superfluous magnificence of buildings. Towers are erected that threaten the stars, excelling the peaks of Parnassus. … Nature complains that she is surpassed by art.…

But he goes on:

> Now the appearance of the building lot is made level with the roller, now the roughness of the surface is vanquished by frequent assault of the builder's ram, now with stakes, thrust into the vitals of the earth, the firmness of the base is tested.[5]

We can see that great attention was paid to preparing the base for the building, not least on making sure that the ground was as even as it could be. Rather charmingly, later on Neckham says that the buildings are so high that if:

> wooden walls have been constructed so proportionally that they are no thicker at the bottom than the top, still the surfaces will not be equidistant. For it is

inevitable that the farther the walls rise from the earth, the greater will the distance between them be found to be.[6]

It seems that he imagined that the posts put into the ground would fan out like the spokes in a wheel because of the curvature of the earth. This might make us smile, but it does show how eye-whackingly high the buildings might seem to the observer at the time. We see cathedrals towering over us and are impressed ... but are we really that impressed, given that we have much higher, much less obviously supported buildings in almost every city? We should not lose sight of the fact that to the medieval man in the street, these buildings were astonishing indeed, and very possibly scratched at the base of Heaven.

Contemporary guides to building a cathedral

There are eight main pieces of writing concerned with how to make things, not just buildings but things connected to building, and it is plain that they were all familiar with the works of Euclid. Occasional other references indicate that the study of geometry was a crucial element of an apprentice's training. The earliest writer was a man who may have been a Benedictine monk, although that is not certain. A seventeenth-century reference puts him as a monk, but elsewhere he was called Rugerus [Roger] and may, therefore, have been the metalworker, Roger of Helmarshausen who was operating around 1100–40. He is also known as Theophilus. Whoever he was, he wrote three books on painting, glass and metalwork. Metal workers and glass artists were crucial to the construction industry and while his treatise does not help the Master Mason, he detailed careful instructions about (for example) how to build the furnace for making glass and how to lay out windows.

> First make yourself a smooth, flat wooden board, wide enough and long enough so you can work two sections of each window on it. Then take a piece of chalk, scrape it with a knife all over the board, sprinkle water on it everywhere and rub it all over with a cloth. When it has dried, take the measurements of one section in a window, and draw it on the board with a rule and compasses. ... After doing this draw as many figures as you wish, first with a [point made of] lead or tin, then with red or black pigment, making all the lines carefully, because, when you have painted the glass, you will have to fit together the shadows and highlights in accordance with [the design on] the board. Then arrange the different kinds of robes and designate the colour of each with a mark in its proper place; and indicate the colour of anything else

you want to paint with a letter. After this, take a lead pot and in it put chalk ground with water. Make yourself two or three brushes out of hair from the tail of a marten, badger, squirrel, or cat or from the mane of a donkey. Now take a piece of glass of whatever kind you have chosen, but larger on all sides than the place in which it is to be set, and lay it on the ground for that place. Then you will see the drawing on the board through the intervening glass, and, following it, draw the outlines only on the glass with chalk. If the glass is so opaque that you cannot see the drawing on the board through it, take a piece of the [clear] white glass and draw it on that. As soon as the chalk is dry, lay the opaque glass over the white glass and hold them up to the light; then draw on the opaque glass in accordance with the lines that you see through it.[7]

He described how to construct not just the forge for metalworkers but the bellows, the anvil, hammers and a variety of tools before he began telling how make chalices, etc and how to apply niello, how to do repoussé work and enamels. Any budding artist would also have found his advice invaluable on what colours and tints have the best effect when painting a young face, an old face, and so on. He did not dictate how to build a cathedral, but his work helps us to understand how craftsmen associated with the building carried out their crafts.

Chronologically, the next major piece of information came from later twelfth-century Canterbury where there had been a disastrous fire in September 1174. The fire had begun in some cottages outside the cathedral gates, but there was such a high wind that sparks soon set fire to the church itself. The account tells how cinders had got caught in the roof beams, but no one realized what was happening until the fire took hold because it was hidden by a painted ceiling. Indeed, people thought that they had done all they could and had gone home without realizing that the cathedral was already beyond help. They did their best to save precious items, some of which were then stolen by the so-called helpers. The congregation's grief was acute. A monk called Gervase wrote a vivid account describing how they wailed and howled their services rather than sang them.[8] He also provided a clear account of what happened next.

William of Sens was the Master Mason who was hired for the rebuild and he set about the task quickly, pulling down the choir, but that was the total work for the first year. Gervase then detailed how William of Sens put up four pillars in 1175, then two more and then three vaulted compartments on each side in 1176. He used quadripartite vaulting with a keystone placed in the middle to lock and unite the parts coming from each side.

In 1177 he set two piers on each side, adorning the two outermost ones [furthest to the east] with engaged columns of marble and making them main piers, since in them the crossing ... and arms of the transept were to meet. After he had set upon these the quadripartite vaults ... with the vaulting, he supplied the lower triforium from the principal tower to the above mentioned piers of the crossing, that is, to the [new western] transept, with many marble columns. Over this triforium he placed another triforium of different material, and the upper windows. Furthermore [he built] the three ribbed vaults of the great vault [of the nave of the choir], namely, from the [old] tower over the crossing to the [new eastern] transept. All of this seemed to us and to all who saw it incomparable and worthy of the highest praise. Joyful, therefore, at this glorious beginning and hopeful to hasten the accomplishment of the work, our hearts were full of fervent longing.[9]

Sadly, this excellent beginning was not to last because William of Sens fell from the scaffolding in 1178, coincidentally five or six hours after a partial eclipse of the sun (which will not have seemed coincidental at the time). He never recovered from his fall and the work passed to William the Englishman in 1179, who had foundations dug to enlarge the eastern end so that a chapel of St Thomas could be built. To do this, William had to dig up the monks' cemetery, so all the bones were carefully collected and reburied in a trench.

As soon as the choir was completed the monks took the new Paschal flame in there on Easter Eve 1180. Gervase's account of this was not just window dressing; Easter that year fell on 20 April, so Easter Eve was St Alphege's Day. Alphege had been Archbishop of Canterbury when he was captured by Viking raiders in 1011 and was killed by them on 19 April 1012 after refusing to allow himself to be ransomed. He was canonized in 1078 and it is thought that Becket had been praying to him just before his own murder. A special resting place had been prepared for Alphege's relics and this was highly symbolic both for the community and healing.

Gervase took a keen interest in the progress of the cathedral, telling us at one point:

If anyone is doubtful as to why the great width of the choir next to the tower should be so much contracted at its head at the end of the church ... the two surviving towers of St Anselm and St Andrew, placed in the circuit of each side of the old church [flanking the passage around the older choir on both sides], would not allow the breadth of the choir to proceed in the direct line. Another reason is, that it was agreed upon and necessary that the chapel of St Thomas should be erected at the head of the church, where the chapel of the Holy Trinity stood, and this was much narrower than the choir. The master, therefore, not choosing to pull down the said towers, and being unable to

move them entire, set out the breadth of the choir in a straight line, as far
as the beginning of the towers ... Then, receding slightly on either side of
the towers, and preserving as much as he could the breadth of the passage
outside the choir [ambulatory] on account of the processions which were then
frequently passing, he gradually and obliquely drew in his work, so that from
opposite the altar ... it might begin to contract, and from thence, at the third
pillar ... might be so narrowed as to coincide with the breadth of the chapel,
which was named of the Holy Trinity. ... All of which may be more clearly
and pleasantly seen by the eyes than taught in writing.[10]

This little excursion shows us some of the problems a Master Mason might have
had to consider. In this case it must have been too difficult or expensive to pull
down the towers, so he had to find a way of using them without compromising
the overall structure. The fact that Gervase wrote at such length about it suggests
that there were numerous comments and discussions. The passage has the air of
a final decision being promulgated.

It is not clear whether or not Villard de Honnecourt of France was a Master
Mason in his own right, or what his precise status was. That he had engineering
knowledge is without doubt; surviving copies of his manuscripts show that he
was the first to make drawings of a new water-powered sawmill operating that
involved two separate motions. He himself says that scientists had long disputed
how to make a wheel turn by itself; his demonstration of a *perpetuum mobile*
is the earliest one found in Europe. He is best known for what remains of his
Portfolio, written in the first half of the thirteenth century, which largely comprises a series of annotated drawings, many of which have geometrical shapes
imposed on them. It is assumed that these were used as teaching aids for young
stone carvers, although he remarks that the discipline of geometry always teaches
one to work more easily. He also includes detailed drawings of buttresses and pinnacles from Rheims cathedral that he appears to have made while the cathedral
was being constructed.

Observe here the rising structure of the chapels of the church at Rheims and
how they are in the interior elevation; the interior passages and the hidden
vaulted-in passages; and on this page you can see the rising structure of the
chapels of the church at Rheims from the exterior from floor to top, just as
they are. ... The first storey of the aisles must form a battlemented parapet so
that a passage can lead around in front of the roof. Against this roof on the
inside are [built] passageways, and, where these passages are vaulted in and
paved, there the passages lead outside, so that one can walk past along the sills
of the windows.[11]

This last sentence reminds us that the Master Masons had to provide access to all parts of the building long after it was completed, for maintenance. Sadly, it seems that a number of Honnecourt's folios have been lost over the years, and it is likely that they were detached on site to help instruct apprentices.

The fourth person on the list is Guilielmus Durandus from Languedoc, also known as William Durand (1230–96). His work, *The Rationale Divinorum Officiorum*, is not quite a manual either. Durandus listed what was equally important to population and Church alike – that is, how to build the required symbolism into the churches. It did not help the Master Mason directly, but it is something in which he will have been interested, so much so that he would probably have followed some of the rules instinctively. For instance, where Durandus says that the foundation must be so contrived that the head of the church must point due east … this would have been taken as read, and still would be to this day. It is rare to find a Christian church built in what used to be known as the Catholic West that does not have its main altar at the eastern end. Occasionally, there are discrepancies where the lie of the land, later building work and usage has led to it being placed differently. Elsewhere, there are variations: Catholic churches in north Africa orientate their churches towards Rome. Durandus gave a lengthy explanation as to why prayer is eastwards and, as this cannot have been written until the second half of the thirteenth century, it might account for an apparent slackness about this up until then. One way to test if a church was originally a much older, Anglo-Saxon establishment is to check how true to east the altar is; if it is off-true, then this may be a clue that it was built on older foundations, which is not uncommon. There is a thought that the earliest method used to orientate the building was to decide to which saint the church was to be dedicated, and then to find east when the sun came up on that saint's feast day (assuming that it did come up). Obviously, there would be a difference between east on the feast of St Mary Magdalene (22 July) and on that of St Andrew (30 November). Equally, there could be local and personal factors to take into account: at what point did one deem the sun to have come up, for example, and who decided?

Durandus assigned symbolic characteristics to the minutiae of the church building. The mortar that holds together the walls is made of lime, sand and water, and, he wrote, represents:

> fervent charity, which joins itself to the sand, that is, undertakings for the temporal welfare of our brethren … the water is an emblem of the spirit. The arrangement of the church resembles that of the human body: the place where the altar is corresponds to the head, the transepts, the arms and hands and the remainder towards the west, the body. The sacrifice at the altar denotes the vows of the heart.[12]

It is possible to see Durandus' ideas in our churches to this day. It should not be thought that Master Masons were building according to Durandus' principles, but more that he was able to find symbolism in almost everything. The four walls should make one think of the four Evangelists. The number eight, revealed in octagonal shapes, symbolizes regeneration, which is why fonts and baptisteries are so often octagonal.[13] All numbers up to fourteen have a meaning and, after that, there may be multiples of numbers. A classic example of number symbolism is the cathedral at Noto, on Sicily, which sits on a hill and is approached by a total of forty-one steps. These are arranged in three flights of eleven plus one of eight at the top. The significance of eight has been mentioned. Eleven is the number of penitence (the number of Apostles left after Judas betrayed Christ), three represents Trinity. Thirty-three is the age Christ is believed to have been when he was crucified. You can make these things say what you want, which was also the point of them; you could use any part of the building to help tell the Christian story.

Durandus has more fanciful measurements such as that the height of the church meant courage, its length, fortitude, and its breadth, charity. Bede had a similar idea (which Durandus undoubtedly borrowed) when he described the dimensions of Monkwearmouth church as having the length of faith, the height of hope, and the breadth of charity. Durandus even has the door representing Christ because sinners enter Christianity through Christ. Elsewhere in his work he gives fascinating insight into how to consecrate a new church, the design of vestments and furniture and how and why certain actions are taken during Mass.

In the fifteenth century, an Italian artist, Cennino Cennini, lived in Florence where he wrote *The Craftsman's Handbook*. This was not dissimilar to Theophilus' work, but Cennini goes further and in more glorious detail. Interestingly, he used paper to trace designs that were to be transferred onto glass. He even instructed would-be craftsmen how to make a quill pens and artists' brushes, and then how you could use the quill pen cuttings to make fine mosaics.

More practical guides to the building process

The last three of the eight medieval authors are writers of manuals as we would understand the term today.[14] Mathes Roriczer, Hanns Schmuttermayer and Lorenz Lechler wrote genuine 'how-to-do-it' books in which they assume a great deal of practical knowledge; one has to examine their drawings closely to understand what is meant. They are technical handbooks, not treatises on theories, which is what makes them so remarkable for that era. They wrote independently of each other. Although all three came from Germany and were contemporaries, roughly speaking, they were usually working quite some distance apart (Roriczer in Regensburg, Schmuttermayer in Nuremberg and Lechler in Heidelberg). Precise dates are not verifiable, but it seems that Roriczer lived from 1430/40– 92/5. Schmuttermayer's career was in full swing in the 1480s so it could be presumed that he was born *c*.1450. The last record of him was in 1518, but this was not a note of his death. Lechler was probably born in 1460, and he died in 1538.

The Roriczer family were the most prolific in that four Master Masons were associated with them in three generations. The senior was Wenzel Roriczer of Regensburg, who died in 1419 leaving his young son, Conrad, to be brought up by his widow and her second husband, the Master Mason Andreas Engel. It seems that Conrad inherited his father's ability and that Andreas was a good teacher; the father and stepson seem to have got on very well and worked together on several projects. Mathes Roriczer was the son of Conrad and it is Mathes who put his thoughts into print. This may have been all the more easy for him because he owned a printing press, although he was not a printer by profession. He was not just a vanity publisher; he produced books by other people as well. Of his own work he published *A Booklet Concerning Pinnacle Correctitude*, *A Booklet on Gables* and *Geometry in German*. It should be mentioned that Roriczer was not trying to teach geometry in *Geometria Deutsch* (they would all have learned it as apprentices). Instead, he offers some useful shortcuts and tips that suggests that pupils then, as now, found some subjects very hard to grasp, even if they were essential to their work. It is much more of a practical application of geometry than the theory that Euclid discussed. Essentially, the *Geometry* was divided into two main area: the first dealt with the construction of simple geometric figures, such as pentagons, heptagons and octagons; the second part focused on solving simple geometric problems, such as how to find the length of the circumference of a circle, how to find the centre of a circle, and how to construct a square and a triangle that have the same areas.

In his *Booklet on Gables*, Roriczer delivered an unprecedented 234-step instruction for making a pinnacle – not a tower or a vault, just a pinnacle. That alone shows how complex the work of the Master Mason was. There are accompanying diagrams in minute detail. He began with a simple square, the corners numbered a–d. This would be the basis for all following measurements, so this is the one that the Master Mason had to get right. Whether Mathes Roriczer wrote the method down from memory, or whether he made some sort of list as he constructed a pinnacle is not known. The plan that he set out was flexible so it could be scaled up or down to make larger or smaller pinnacles. Throughout his instructions, he continually advises that a Master Mason must set his dividers to the precise width required very carefully because all other measurements will follow from that one. A Master Mason was always depicted with dividers and a

set square, as was God when portrayed as God the Architect, the Creator of the
World; if there is an effigy on a Master Mason's tomb, the same tools are usually
displayed. Everything depended on the accuracy of that first measurement. Once,
I was interested to talk to a stone mason at Durham cathedral who told me that
he always relied on measurements made with dividers because, perhaps surprisingly, laser measurements are not always as accurate.

It has sometimes been suggested that Mathes Roriczer was nowhere near the
measure of his father, stepfather or grandfather in terms of construction skill,
but we are indebted to him for giving such insight into how they set about the
business. It was for this work that he was accused of breaking his vow of secrecy.
The outcome of the accusation is not known, but it might explain why he wrote
a preface to his booklet on pinnacles in which he says that the bishop of Eichstätt
had thought it would be a good idea. His thought, apparently, was that those who
were professional masons would benefit from it and those who did not understand
their craft would be spotted and got rid of. That stands up to a certain point, but,
as we have seen in Chapter 6, regulated training overseen by guilds was already
in place. Schmuttermayer wrote that he had produced his booklet to bring it:

> into the most understandable [form] by means of setting it out, neither with
> too little description nor with more words than is necessary.[15]

He then credits well-known Master Masons with bringing the techniques to
light. He said that he was asked to write the booklet so that craftsmen could
improve and to help them to memorize the methods used.

Hanns Schmuttermayer was probably a goldsmith with a special interest in
minting coins. If he was, there is no explanation as to why he would describe
how to design pinnacles other than that many specialists also had detailed
general knowledge. In 1487, he was sued by a woman who claimed that he had a
silver cane that belonged to her, which does not place him in any specific trade,
but snippets of information suggest that he was a goldsmith. One record says
that someone owed money both to him and to Albrecht Dürer (it is thought
that this is a reference to Albrecht senior, who was a goldsmith, not Albrecht
junior, the artist). More usefully there is a 1489 agreement recorded between
'Schmuttermayer, goldsmith, his wife Agnes' and some cousins concerning standing as surety for some money.[16] In 1503, he was permitted to be mint-warden for
a year at Schwabach, and he may have held that position longer.

Lon Shelby has examined Roriczer's and Schmuttermayer's works in detail
and reports that Schmuttermayer's is a booklet of only six folios with text on
both sides of four of them; one of the reverse folios is blank while the other has a
geometric setting out of a pinnacle and a gablet, so there is sadly little to go on.

There is, however, a chance that Schmuttermayer produced his booklet first and that is what inspired Roriczer rather than the other way around as might have been expected. His is also a step-by-step guide beginning:

> In the name of our Lord, Amen. If you wish to draw a pinnacle and a gablet, then first make a square, however large you wish. In the same square make eight squares smaller and smaller, so that each fits into the other on the diagonal … Then set the eight squares all equally beside one another and give to each a letter …[17]

Lorenz Lechler was a late fifteenth-century German Master Mason who wrote a booklet on Gothic design called *Instructions*, and who is known to have worked on Heidelberg church as well as on that of St Dionysius in Esslingen. He wrote from the point of view of a practising mason and specifically for his son, Moritz who, we presume, was a fledgling Master Mason. He used a ground plan by way of illustration and explanation, but the text is not as systematic as those of Roriczer and Schmuttermayer. Lechler also based his description firmly in the practical application of geometry, showing how the width of a church choir often became the basic unit from which all other measurements ensued. For example, the outside wall of the church should be one-tenth the width of the choir, and that measurement should be used to generate smaller dimensions for buttresses, windows, and so on, so that everything would appear in proportion. Perhaps alarmingly – or charmingly – he noted that an 'honourable work glorifies its master, if it stands up'.

As time went on, and with the Renaissance in full swing, especially in Italy, many books were produced, but the ones mentioned above were the ground-breakers. Of course, it is more than possible that there were many others; the fact that only fragments remain of some documents attests to the fact that a great deal has been lost to us. For instance, John Shute wrote *The First and Chief Grounds of Architecture* in 1563 after he had visited Italy.[18] He was a painter and architect (his description), and he has the distinction of being the first Englishman on record to use the word 'architecture'. He had read Vitruvius and almost anything else that he could get his hands on whether in Latin, Italian, French or Dutch, after which he opted to go for a practical approach. He was clear that the Renaissance styles had to be learned, almost as a language must be, but he did lay out measurements and proportions needed for would-be architects to achieve their aim.

The language about which Shute was so passionate began to cause problems in itself. No one who is not part of a team likes jargon, which proved to be the same in the sixteenth century. Bertrand Jestaz found the following delightful passage in Noël du Fail's *Contes et discours d'Eutrapel* of 1585:

One day Pihourt, a mason of Rennes, saddled up his mare and, with straw on his boots, a robe roughly tied with string around the waist and his hat askew, rode to Chateaubriant where a fine chateau was to be built. When he heard the great craftsmen who had been summoned from all over France talking of nothing but frontispieces, pedestals, obelisks, columns, capitals, friezes, cornices, dados, none of which he had ever heard of in his life, he was dumbfounded. When it was his time to speak … he said … that in his opinion the building should be built well and competently, with an adequate *palison* as required. When he had had his say, all those present deemed him to be a great man, to be listened to with care on this important matter that they were unable fully to comprehend; they reckoned that he must know a thing or two. Before he went away, the victorious builder warned that he could stay no longer but that the set of the sleeves could never be correct without him and the equipolation of his heteroclites. The assembled company were amazed by him (they had no idea what he was talking about), and this gave rise to the saying: 'as resolute as Pihourt and his heteroclites'.[19]

Clues as to how Master Masons physically built can also be found in sometimes ancient ruins. At Segesta, on Sicily, and at Delphi, in Greece, stonework has not been finished, so lumps of stone can be seen sticking out that have not been cut back. Ropes were tied round these protruding lumps to help hoist them into position. Once laid, they would normally have been cut back to present a smooth edge, but something happened to prevent that. Elsewhere on Sicily, in the Valley of the Temples, there are stones that have grooves to accommodate lifting ropes. We also know that scaffolding techniques improved so that scaffolding could be suspended from above, not just built up from the ground. To this day, the scaffolding marks remain on numerous buildings in the form of putlogs; distinctive, regularly set holes in walls where scaffolding was placed (*see* plate 16). They can also indicate where beams supported upper floors, so a ruin should be examined with that thought in mind. Presumably, they were originally filled in with plaster to make a more pleasing finish.

Other contemporary sources of the building process

We get more information from the numerous images of building sites in manuscripts, occasional wall paintings and carvings. For example, in a fresco painted for the Ospedale di Santa Maria della Scala, in Siena, in 1443, Domenico di Barrolo shows a bishop distributing alms in a piazza where builders are working. The man with dividers in the front seems to be taking measurements from a plan, another is picking up bricks, a third climbing a ladder with a hod on his shoulder. A bucket of cement is being carried by pulley up onto the scaffolding. In spite of such realistic touches, the buildings in the background are structurally unrealistic, but that does not matter: the point of the image is to show the largesse of the bishop – the building project in the background may well be incidental, although it usually indicates that the central figure has also contributed to the building, or has inspired others to do so.

It is not unknown to find carvings of building sites. The façade of Nîmes cathedral (*see* plate 17) shows masons at work while a man operates a hoist from a diminutive tower on the left. Venice has some charming capitals in St Mark's Square, which show sculptors at work (*see* plate 18).

Nearer to home, we see Matthew Paris' mid-thirteenth-century drawing of Offa with his Master Mason and workers on St Alban's site. While the Master Mason is clearly explaining things (he has one hand raised with a pointing finger in the gesture that means that he is saying something), other figures are busy: two men carry stones or tiles up a laddered ramp on a stretcher while another pushes a laden wheelbarrow. On the opposite page of Paris' *Life of Saint Alban and Saint Amphibalus*, there is a vignette of six workers.: one is using a pulley to lift a basket of stones; one uses a level to make sure a wall is even while a third figure balances precariously, using a plumbline; three other men work on a ladder, at a small anvil and at a trestle table, a variety of tools around them such as a mallet, a broad axe and a stone hammer. The table is of the collapsible sort seen endlessly in medieval manuscripts and carvings, especially in dining scenes. So widespread were these images of builders and building sites right across Europe that Günter Binding has been able to produce line drawings of 673 of them from the early ninth century to *c.*1500 in his book, *Medieval Building Techniques*. He identified at least sixty-five different tools in regular use.

As well as stone, it is obvious that wood, metal and glass were crucial ingredients of a cathedral, so the men (and, occasionally, women) who worked in those areas were also a vital element of how a cathedral was constructed, the Master Mason having to coordinate all parties. Carpenters and carvers have been mentioned from time to time, not least the fact that approximately a third of Master Masons began life in those professions. Any worker in wood was vital whether they were the sawyers who turned felled trees into planks or they made wheels for carts, tile pins or the frames that were often the basis from which so much else depended. Although it is possible to find records of specific trades such as sawyer, joiner, cartwright and so on, in practice they could probably all turn their hands to each other's specialities. It is often remarked that the inside of church and cathedral roofs look like upside-down ships and that the word 'nave' derives from the Latin *navis* ('a ship'). Men who could produce watertight, shaped structures that could withstand the battering of the weather were in demand and many of these were shipwrights. Church guide books sometimes put forward the idea that the timbers for a roof were recycled from wrecked ships, and there may well be truth in that: not only would some of the shapes be useful, but the sea water would have helped to season them, too.

Dead wood was mainly used for making charcoal, which had several uses in the Middle Ages, not least by artists sketching out designs. On the building site, charcoal was useful because it could reach higher and more sustained temperatures than coal and wood, and was the only fuel used for iron smelting until coal was first successfully introduced in 1620.[20] It was a surprisingly lucrative business, but one that took its toll on the countryside; even in the fourteenth century, people were worried about deforestation, although their primary concern was to safeguard future business. In 1302 it was decided that if a wood was sold to colliers, the animals there would have no cover 'on account of the fire and noise made by the charcoal burners, and by reason of the destruction of the oaks and other trees'. The impact on the charcoal burner was that they could not work unless they had a licence and that they could end up in court if they operated without one. As charcoal burned very evenly, it was useful to smelters of bronze and iron as well as to glass-makers.

The smelters, of course, were essential to the smiths and the smiths were vital to the cathedrals. Described in a mid-fourteenth-century satire as 'Swart smoky smiths smirched with smoke', smiths were largely using any metal on which they could lay their hands for a good price, so there was a lot of recycling of scrap metal. As with wood-workers, there were many different specialist trades within the general term of smith: chain-makers, lorimers, bladesmiths, those who made armour, cutlery, door hinges, buckles and buckets and locks, and a smith's skills as a farrier would have been far more in demand then than now. Interestingly, the Holkham Bible, made in the fourteenth century, shows a woman blacksmith. This may not be a social comment on equality so much as an idea coming out of medieval literature that a female blacksmith was particularly abhorrent, taking over the job of making the nails that would fix Christ to the cross after her husband pretended to have hurt his hand to get out of doing it.[21] However, there are accounts of real women working in forges: Katherine of Bury was the widow and the mother of a king's smith, and she was retained at the sum of 8d a day to keep the forge going in the Tower of London while her son was away campaigning at Crécy. There are other women occasionally mentioned in accounts of other trades, usually in a helping-out role and usually paid rather less than the men.

One imagines that there might be something solid and straightforward about blacksmiths. Nonetheless, the Master Mason needed to have been aware that blacksmiths, too, had their foibles. Many refused to shoe a horse on the feast day of St Eligius (1 December). Eligius (also known as Eloi) was the royal goldsmith and minter at the seventh-century Frankish court. Having given generously to many monasteries and churches, he gave up smithing to become a priest and later was bishop of Noyon. He was known for his forbearance: to swear by St Eloi (as did Chaucer's Prioress) meant not to swear at all because he never did, despite the number of times he must have hit his thumb.

Goldsmiths were a breed apart. Alexander Neckham, whom we met earlier in this chapter, gave this description of goldsmiths working in Paris in the late twelfth century:

The goldsmith should have a furnace with a hole at the top so that the smoke can get out. One hand should govern the bellows with light pressure and with the greatest care so that the air pressed through the nozzle may blow upon the coals and feed the fire. Let him have an anvil of extreme hardness on which the iron or gold may be laid and softened and may take the required form. They can be stretched and pulled with the tongs and the hammer. There should also be a hammer for making gold leaf, as well as sheets of silver, tin, brass, iron or copper.

The goldsmith must have a very sharp chisel with which he can engrave figures of many kinds on amber, hard stone, marble, emerald, sapphire or pearl. He should have a touchstone for testing and distinguishing between iron and steel. He must also have a rabbit's foot for smoothing, polishing and wiping the surface of the gold and silver. The small particles of metal should be collected in a leather apron.

He must have small pottery vessels and cruets, and a toothed saw and file for gold as well as gold and silver wire with which broken objects can be mended. He must also be as skilled in engraving as well as in bas relief, in casting and as well as in hammering. His apprentice must have a waxed table, or one covered with clay, for portraying little flowers and drawing in various ways.

He must know how to distinguish pure gold from latten and copper, lest he buy latten for pure gold. For it is difficult to escape wiliness of the fraudulent merchant.[22]

In 1292 a tax list was drawn up for Paris, which listed 116 goldsmiths (possibly all one needs to know about the standard of living at that time). Not all of them would have been employed on gold-work for the Church; many would have fashioned personal jewellery as well.

Glaziers

Coloured glass has existed in Britain since Roman times. Roman examples can be seen in the British Museum and elsewhere, so the concept was nothing new. There is *millefiori* glass on some of the treasures buried within a ship at Sutton Hoo around the year 615. We also know that coloured window-glass arrived here courtesy of Benedict Biscop at the end of the seventh century, when he brought glaziers and others with him on his return from one of his trips to Rome. Certainly, the Vikings and Anglo-Saxons loved coloured glass beads as part of their jewellery. Six hundred years later, William of Malmesbury described Canterbury cathedral's west end as 'a blaze of glass windows', so glass was certainly being used to magnificent effect in cathedrals.

The first time glass was produced on a large scale in England was in 1226 at Chiddingfold, Surrey. There, a Norman called Laurence Vitrearius set up what seems to have been a permanent workshop in that it was not directly dependent upon a local building project. That general area was an obvious centre for production because of the seemingly inexhaustible supply of beech wood to provide the potash flux. The production method was of a cylinder-blow variety, known as broad sheet. In 1240, Vitrearius's name occurs in association with the supply of glass to Henry III's new building at Westminster Abbey, and by the time of Elizabeth I's reign, there were at least eleven glass works on Chiddingfold's green. So fine was the broad sheet glass that it was used in some of the most prestigious buildings in the land, such as the chapels of St Stephen at Westminster and St George at Windsor, although there was still a brisk business in importing it. In Exeter cathedral, glass dating to 1317–18 came from Rouen. Transporting glass was expensive and then, once it had arrived, lead had to be obtained to set it within the window. Intriguingly, another glazing expense was for ale bought for whitening the glaziers' tables. This was because designs were marked out on tables that had to be white for the drawings to be clear. A solution of chalk and water was rubbed into the table surface, but ale worked better because the sugar in it made the chalk bind better. The outline pattern could later be wiped off and the table-top reused. It worked as a horizontal whiteboard. Instructions for transferring designs were in circulation from Theophilus and Cennini, and doubtless from others too.

One does not often find references to maintenance, but glass was and is both fragile and expensive, so it had to be looked after. In 1240, an arrangement was made by Chichester cathedral with John the Glazier: in return for a daily allowance of bread and a yearly fee of 13s 4d, John and his heirs were to keep the windows of the cathedral in good repair:

They shall preserve the ancient glass, and what has to be washed and cleaned they shall wash and clean, and what has to be repaired they shall repair, at the cost of the church, and what has to be added, they shall add, likewise at the cost of the church; and there shall be allowed them for each foot of addition one penny. And as often as they repair the glass windows, in whole or in part, they shall be bound, if so ordered by the keeper of the works, to make one roundel with an image in each. And if they make a new glass window entirely at their own cost, which is without pictorial decoration and is fifty-three feet in total area they shall receive for it and their expense.[23]

The mention of John the Glazier's heirs indicates that the cathedral Chapter saw this as a long-term investment. Interestingly, the 53ft (16m) referred to in the agreement corresponds to the dimensions of the contemporary clerestory windows at Chichester.

These records and accounts give us much more than just insight into how cathedrals were built, but how individual trades were inextricably linked with others. To take but one example, we can see how the whole project hung together by looking at the accounts for Exeter cathedral. Detailed accounts survive for the years 1279–1353, of which almost any page, opened at random, offers illustrations of people working on the site. In week four of the Michaelmas Term 1329, the following was paid out in wages:[24]

And to Yeul the roofer with his servant 2s 6d over the new lodge, and Breche with his servant for one day 5d, and Breche's servant for 2 days 4d to a small boy for 2 days 3d.
4,000 stone-pins bought 4d.
2,000 'latnall' bought 20d.
In wages of 6 servants in the church 5s 10d each.
In 2 carters 23d.
In 3 labourers in the quarry 2s 9d, 11d each.
In 5 labourers in the quarry 4s 10d each.
2 horse-collars bought 21d.
One 'lordrop' and 8 pairs of traces, 2 cart ropes and 6 'tainsefftes' [ropes] bought, weighing 7 stones, each stone of 20 lb weight, 9s 4d, price 16d a stone.
2 pieces of 'wyppyn-cord' bought 4d.
For the fodder for the horse which procured the same at Bridport 2d.
One horse-hide bought 15d.
18 pairs of hinges 15 pairs of hooks bought for the new lodge 3s 1d.
For steeling the masons' tools 3s 11d.

11½ quarters of lime bought 4s 9½d 5d a quarter. Fodder as above
Total: £4 12s 5d

This list reveals men labouring in the quarry, materials being transported, equipment needed to keep everything moving and to keep the administration of the lodge rolling along, even small children being employed on that behemoth, the cathedral building site. One can imagine the diverse characters of the people mentioned, some accepting of their lot, others demanding more pay and better conditions. Perhaps we can imagine them standing back from the cathedral, gazing ever upwards and feeling immense pride that they had been part of it, knowing that they must have helped improve their chances of getting into Heaven. All these men and some women were controlled and coordinated by the Master Mason, so it is understandable if he were sometimes aware of his position and own importance.

AFTERWORD

In the Middle Ages, as now, building was a big industry, and there was much to do in the administering and monitoring of it, giving employment to thousands. There was a race, not just for the sky, but for winning new business. There were new shrines to be built, more and more cathedrals to be adapted, as fast as possible, not only to accommodate the demand, but to capitalise on that demand. That should not detract from the main reason for building them, which was to give glory to God. The main function was, and still is, to worship and perpetuate the Christian message. Every now and then a journalist will write an excited piece about a cathedral or perhaps several, describing them as forgotten masterpieces, but that says more about the writer than the cathedral. In fact, they are one of the most enduring and durable features of our landscape. Right across Britain and Europe and beyond, we can see the same aim, but sometimes taking very different forms, whether in the long, low, elegant English versions, the dizzyingly high French ones, the stately, ornate Spanish ones, or the round-domed Italian. Each one has been created with maximum concentration and passion to be the greatest masterpiece, not just of that city, but of the era.

There is nothing wrong with simply visiting a cathedral today as a tourist and enjoying its many works of art, the surprising elements of humour and local history entombed within, and noting the minute detail inside and out. In this book, I have tried to seek out the people who made these buildings, and it is not necessarily true that they all worked anonymously through choice. They were human beings and, of course, they wanted recognition, which is why so many of their names survive, whether through accounts, contracts, wills, or stylistic signatures. We would have liked them to have been much more specific, but perhaps more information would rob the buildings of some of their mystique? Nonetheless, we are able to see the builders as real people. Cathedral building was such a huge part of medieval daily life that we should not be surprised by the large number of accounts that survive. Not everyone would have been overjoyed when the building circus came to town, especially if it involved the removal of their property to make way for the building site. As soon as the idea of a new church was mooted it seems that the area became flooded by Master Masons putting forward plans and, more annoyingly, by migrant workers. This was a much more itinerant age than is commonly imagined; the roads were not only filled by pilgrims, but by what must have seemed then to have been an endless river of humanity looking for work. A new-build cathedral will have attracted the homeless, hopeless and highly skilled, all looking for shelter, food, clothing and pay.

It is almost beyond our imagination to grasp how the coordination of it all worked. How could the Master Mason take on such a gigantic project with such confidence? How could he be sure that the quarry would provide enough stone and that he would find enough timber and overcome all the other logistical difficulties? It is very clear that the Master Masons were exceptional men in terms of leadership and, like all the best leaders, they understood that management of manpower and materials was as critical as understanding how to get the project off the ground. They had to keep the motivation going and satisfy the often grandiose ideas of the patrons with their unrealistic timeframes. Some Master Masons became very wealthy, but they earned it. Through their works, we can look back and see real people working together, becoming friends, laughing and sometimes crying together when death and disease all too frequently struck. We see then the outcome of so many people at so many levels, stages and intent interacting with the buildings. We can almost see their footprints. We are looking at the passion and heart's blood of the Master Masons, patrons and artists, some of whom perished in the attempt to achieve their dream to recreate Heaven on earth. In their creations we see professional men who were tested in long apprenticeships and who took their work seriously, not just for their own reputation, but for God's. The critical factor that made the buildings so extraordinary and which gives them the star quality we now seek in lesser things, is belief. These men believed, not slavishly but with understanding, fear and hope for their own eternal future. Their aim was not so far, after all, from wandering hermits seeking Paradise on earth. The difference was that the wanderers went looking for it; the Master Masons created it.

Thy Will be done on Earth as it is in Heaven.

APPENDIX 1

TRADES ASSOCIATED WITH A MEDIEVAL BUILDING SITE

Alabasterer
Apparitor (foreman)
Artist (for murals)
Bell founder
Blacksmith
Bladesmith

Boarder
Boatman
Brassworker
Bricklayer
Bronze caster
Carpenter
Carter
Cartwright
Carver
Charcoal burner
Clerk
Cook
Cutler
Dauber
Digger

Ditcher
Docker
Earthwaller
Embroiderer
Engineer
Farrier
Forester
Free-stone mason*

Furbisher
Gardener
Geometer
Glazier
Goldsmith
Hardhewer (dressed a variety of
    Kent hard stone)
Hewer
Hodman
Imager
Iron worker
Joiner
Labourer
Layer
Lime burner
Locksmith
Lorimer
Marbler
Mason / Cementarius
Millwright
Mortarer
Ostler / stable-boy

Ox driver
Painter
Paper maker
Pargetter / Plasterer
Parlier (translator / right-hand man)
Paviour / Paver
Plumber
Polisher (of marble)

Portehache (tool carrier)
Poser (positioner of stones)
Quarryman
Roofer
Rope maker
Sawyer

Scaffolder
Scappler (cut stone roughly to
    shape)
Sculptor
Silversmith
Slater
Stone-cutter
Strawman (wattle-and-daub)
Surveyor
Tanner
Tapicer (maker of carpet /
    tapestries)
Thatcher
Tiler
Tomb maker
Tree splitter
Upholder (maker of props, also an
    undertaker)
Vat maker
Waller
Wax chandler
Weaver
Wheelwright
Whitewasher

*   Free-stone is the name given to a fine-grained sandstone or limestone that can be cut or sawn in any direction. It can be undercut and so used for carvings in relief and it can be dressed into practically any geometrical shape. A free-stone mason was both an artist, in the sense of being a carver, and a precision worker in that he often had to cut exact parts to be fitted together for windows, fan-vaults and so on

APPENDIX 2

BREAKDOWN OF MASTER MASONS' ORIGINAL TRADES

Trade	Number	Percentage of individuals name in records (AD 968–1540)
Mason	693	54
Carpenter	411	32
Carver	77	6
Engineer	19	1.48
Joiner	16	1.3
Marbler	12	0.9
Mason/Carver	7	0.55
Sculptor	4	0.31
Carpenter/Engineer	3	0.23
Clerk	3	0.23
Imager	3	0.23
Carpenter/Carver	3	0.23
Bricklayer	2	0.155
Brick-maker	2	0.155
Mason/Sculptor	2	0.155
Carver/Joiner	2	0.155
Carpenter/Timbermonger	2	0.155
Millwright	2	0.155
Architect	1	0.08
Brassworker	1	0.08
Carpenter/Joiner	1	0.08
Engineer/Mason	1	0.08
Goldsmith	1	0.08
Layman	1	0.08
Mason/Tiler	1	0.08
Mason/Carpenter	1	0.08
Mason/Marbler	1	0.08
Tomb-maker	1	0.08

NOTES

Chapter One: New technology or new spirituality?

1. See, for example, 1 Kings 6:2–10 and Ezekiel 40:5–47 and 43:13–17 (King James's Version).
2. Harvey, J., *English Mediaeval Architects: A Biographical Dictionary down to 1550* (Batsford, 1954), p.156.
3. Erlande-Brandenburg, A., *Cathedrals and Castles: Building in the Middle Ages*, (Harry N. Abrams, 1995), p.121.
4. Harvey, *English Mediaeval Architects* (Batsford Ltd, 1954), pp.199 & 212.
5. Ashlar is masonry made up of accurately squared stones laid in regular courses with a smooth face and fine joints.
6. Grant, L., *Abbot Suger of St-Denis* (Longman, 1998), p.29.
7. Abbot Suger, *De Administratione*, Burr, D. (trans.), Chapter XXV (1996) in Frisch, T. G. *Gothic Art 1140–c. 1450: sources and documents* (University of Toronto Press, 1971), p.6.
8. Abbot Suger, *De Administratione*, Burr, D. (trans.), Chapter XXV (1996) in Frisch, *Gothic Art 1140–c. 1450*, p.7
9. Cross, F.L. (ed.), *The Oxford Dictionary of the Christian Church* (Oxford University Press, 1958), p.699.
10. Gameson, R.G.G., 'thelwold, Benedictional of', in *The Blackwell Encyclopaedia of Anglo-Saxon England* (Blackwell, 2001), p.20.

Chapter Two: Paving the way

1. Icher, F., *Building the Great Cathedrals* (Harry N. Abrams, 1998), p.50.
2. Frisch, *Gothic Art 1140–c. 1450: Sources and Documents*, p.17.
3. Icher, *Building the Great Cathedrals*, p.176.
4. Harvey, J., *English Mediaeval Architects: A Biographical Dictionary down to 1550* (Batsford, 1954), pp.31 & 206.
5. Lucie-Smith, E., *The Thames and Hudson Dictionary of Art Terms* (Thames & Hudson, 1996), p.90.

6.   Salzman, L. F., *Building in England down to 1540: A Documentary History* (Clarendon Press, 1952), p.383.

7.   Salzman, *Building in England down to 1540*, p.364.

8.   Erlande-Brandenburg, A. *Cathedrals and Castles: Building in the Middle Ages*, p.135.

9.   Salzman, *Building in England down to 1540*, p. 359.

10.   Gimpel, J., *The Cathedral Builders* (Grove Press, 1961).

11.   Coulton, G. C. *Life in the Middle Ages, Volume 1* (Cambridge University Press, 1930), p.3.

12.   Harvey, *English Mediaeval Architects*, p.124.

13.   Harvey, *English Mediaeval Architects*, p. 136.

14.   Knoop, D. & Jones, G.P., *The Mediaeval Mason, an Economic History of English Stone Building in the Later Middle Ages and Early Modern Times* (University of Manchester Press, 1933), p.47.

15.   Scott, J. & Gray, G., *Out of the Darkness* (Axminster Printing, undated), pp.1 & 17.

16.   Frisch, T. G., *Gothic Art 1140–c.1450: Sources and Documents* (University of Toronto Press, 1987), p.94.

17.   Icher, *Building the Great Cathedrals*, p.80.

18.   Abbot Suger, *De Administratione*, Burr, D. (trans.), Chapter XXV (1996) in Erlande-Brandenburg, *Cathedrals and Castles*, pp. 141–2.

19.   Frisch, *Gothic Art 1140–c.1450*, p.29.

20.   Frisch, *Gothic Art 1140–c. 1450*, p.26.

21.   Burckhardt, T., *Chartres and the Birth of the Cathedral* (Golgonooza Press, 1995), p.84.

22.   Salzman, *Building in England down to 1540*, pp.132–36.

23.   Salzman, *Building in England down to 1540*, p.120.

24.   Harvey, *English Mediaeval Architects*, p.298.

25.   Harvey, *English Mediaeval Architects*, p.239.

26.   Cannon, J., *Cathedral: The Great English Cathedrals and the World that made them* (Constable & Robinson, 2007), pp.484–85.

27.   Crowland Abbey Guide, 2007, p.4.

28.   Frisch, *Gothic Art 1140–c.1450*, p.27.

29.   Salzman, *Building in England down to 1540*, pp.364–5.

30.   Icher, *Building the Great Cathedrals*, p.45

31.   Brown, M. P., *The World of the Luttrell Psalter* (British Library, 2006), pp.24–5.

32.   Ziegler, P., *The Black Death* (Sutton Publishing, 1997), p.51.

33.   Chaucer, G., *The Canterbury Tales*, Coghill, N. (trans.), (Penguin, 1960), lines 691–703.

34.   Salzman, *Building in England down to 1540*, p.376.

35.   Icher, *Building the Great Cathedrals*, pp.169.

Chapter Three: The Master Masons I

1.   Bowyer-Smyth, V., *From Caen to Canterbury: The Stone in the Cathedral* (undated pamphlet), p.7.

2.   Knoop, D. & Jones, J. P., *The Mediaeval Mason, an Economic History of English Stone*

Building in the Later Middle Ages and Early Modern Times (University of Manchester Press, 1933), p.65.

3. Erlande-Brandenburg, A., *Cathedrals and Castles: Building in the Middle Ages* (Harry N. Abrams, 1995), p.102.

4. Frisch, T. G., *Gothic Art 1140–c.1450: Sources and Documents* (University of Toronto Press, 1987), p.94.

5. Cannon, J., *Cathedral* (Constable & Robinson), p.222.

6. Lethaby, W. R., *Westminster Abbey: The King's Craftsmen* (Duckworth & Co., 1956) p.158.

7. Harvey, J., *English Mediaeval Architects: a Biographical Dictionary down to 1550* (Batsford, 1954), p.11.

8. Frisch, *Gothic Art 1140–c.1450*, p.80.

9. Phillips, J., 'Beverley Minster nave: the evidence of the masons' marks', P.S. Barnwell & A. Pacey (eds), *Who Built Beverley Minster?* (Spire Books, 2008), p.41.

10. Icher, F., *Building the Great Cathedrals* (Harry N. Abrams, 1998), pp.162–85.

11. Alexander, J. S., 'Masons' marks and the working practices of medieval stone masons', in Barnwell & Pacey, p.23.

12. Alexander, J. S., 'Masons' marks and the working practices of medieval stone masons', in Barnwell & Pacey, p.24.

13. Plagnieux, P., *Amiens, the Cathedral of Notre-Dame* (Monum Editions, 2005), p.45.

14. Lysons, D., & Lysons, S., 'Antiquities: Ancient church architecture', *Magna Britannia: volume 4: Cumberland* (1816), pp.189–202.

15. Anderson, M.D., *History and Imagery in British Churches* (John Murray, 1971), p.61.

16. Butler, A., *Lives of the Saints* (Burns & Oates, 1985), p.17.

17. Harvey, *English Mediaeval Architects*, p.70.

18. Harvey, *English Mediaeval Architects*, p.23.

19. Harvey, *English Mediaeval Architects*, p.172.

20. Harvey, *English Mediaeval Architects*, p.172.

21. Harvey, *English Mediaeval Architects*, p.43.

22. Harvey, *English Mediaeval Architects*, p.220.

23. Dyer, C., *Standards of Living in the Later Middle Ages* (Yale University Press, 2002), p.249.

24. Harvey, *English Mediaeval Architects*, p.20.

Chapter Four: The Master Masons II

1. Salzman, L. F., *Building in England down to 1540: A Documentary History* (Clarendon Press, 1952), p.56.

2. Sharpe, R. (ed.), 'Folios ccxxii–ccxxx: Items relating to 1331–1334', *Calendar of Letter-books of the City of London*: E: 1314–1337 (His Majesty's Stationery Office, 1903), pp.262–71.

3. Goldberg, P. J. P., *Women in England c.1275–1525* (Manchester University Press, 1995), p.10.

4.   Goldberg, P. J. P., *Women, Work, and Life Cycle in a Medieval Economy* (Clarendon Press, 1992), p.328.

5.   Goldberg, *Women in England c.1275–1525*, p.6.

6.   Erlande-Brandenburg, A., *Cathedrals and Castles: Building in the Middle Ages* (Harry N. Abrams, 1995), p.134.

7.   Erlande-Brandenburg, A., *Cathedrals and Castles*, p.144.

8.   Harvey, J., *English Mediaeval Architects: A Biographical Dictionary down to 1550* (Batsford, 1954), p.169.

9.   Frisch, T. G., *Gothic Art 1140–c.1450: Sources and Documents* (University of Toronto Press, 1987), p.55

10. Harvey, *English Mediaeval Architects*, p.37.

11. Harvey, *English Mediaeval Architects*, p.222.

12. Harvey, *English Mediaeval Architects*, p.21.

13. Harvey, *English Mediaeval Architects*, p.177.

14. Benedict, St., *The Rule of St Benedict*, Abbot Parry OSB (trans.) (Gracewing, 1990)

15. Harvey, *English Mediaeval Architects*, p.71, p.83

16. Acts 5:1–11 (King James's Version).

17. Gimpel, J., *The Cathedral Builders* (Grove Press, 1961), p.104.

18. Harvey, *English Mediaeval Architects*, p.74.

19. Jones, K., *Gender and Petty Crime in Late Medieval England* (Boydell Press, 2006), p.112.

20. Butcher, A.F., 'The Origins of Romney Freemen, 1433–1512', *Economic History Review*, Vol. 27, p.25.

21. Myers, A.R. (ed.), *English Historical Documents*, Vol. IV (Eyre & Spottiswood, 1969), pp.569–70.

22. Swanson, H., *Building Craftsmen in Late Medieval York* (University of York, 1983), p.28.

23. Harvey, *English Mediaeval Architects*, p.172.

24. Harvey, *English Mediaeval Architects*, p.126.

25. Harvey, *English Mediaeval Architects*, p.243.

26. Dyer, C., *Standards of Living in the Later Middle Ages* (Yale University Press, 2002), p.209.

27. Jusserand, J. J., *English Wayfaring Life in the XIVth Century* (first published 1889; reprinted Kessinger Publishing, 2003), pp.116–17.

Chapter Five: Building controls

1.   Blair, J., & Ramsey, N. (eds.), *English Medieval Industries* (Hambledon, 2001), p.65.

2.   Watson, P., *Building the Medieval Cathedrals* (Cambridge University Press, 1976), p.37.

3.   Salzman, L. F., *Building in England down to 1540: A Documentary History* (Clarendon Press, 1952), p.376.

4.   Salzman, *Building in England down to 1540,* p.377.

5.   Salzman, *Building in England down to 1540,* p.378.

6.   Salzman, *Building in England down to 1540,* p.377.

7.   Salzman, *Building in England down to 1540*, p.363.
8.   Erlande-Brandenburg, A., *Cathedrals and Castles: Building in the Middle Ages*, (Harry N. Abrams, 1995), p.153.
9.   Senderowitz Loengard, J. (ed.), 'Introduction', *London viewers and their certificates, 1508–1558: Certificates of the sworn viewers of the City of London* (1989), p.11.
10.   Salzman, *Building in England down to 1540*, p.368.
11.   Swanson, H., *Building Craftsmen in Late Medieval York* (University of York, 1983), p.16.
12.   A mileway was the time needed to walk a mile.
13.   St Helen's Mass is 3 May and Lammas is 1 August. As St Helen's day is 18 August, the day referred to is the Invention of the Holy Cross, discovered by St Helen.
14.   Myers, A. R. (ed.), *English Historical Documents 1327–1485*, Vol. IV, p.1087.
15.   Frisch, *Gothic Art 1140–c.1450*, p.55.
16.   Frisch, *Gothic Art 1140–c. 1450*, p.104.
17.   Icher, F., *Building the Great Cathedrals* (Harry N. Abrams, 1998), p.171.
18.   Frisch, *Gothic Art 1140–c. 1450*, p.55.
19.   Frisch, *Gothic Art 1140–c. 1450*, p.54.

Chapter Six: Training and beyond

1.   Icher. F., *Building the Great Cathedrals* (Harry N. Abrams, 1998), p.184.
2.   Icher, *Building the Great Cathedrals*, p.163.
3.   Icher, *Building the Great Cathedrals*, p.184.
4.   Icher, *Building the Great Cathedrals*, p.182.
5.   Icher, *Building the Great Cathedrals*, p.164.
6.   Icher, *Building the Great Cathedrals*, p.185.
7.   Icher, *Building the Great Cathedrals*, p.185.
8.   Icher, *Building the Great Cathedrals*, p.182.
9.   Icher, *Building the Great Cathedrals*, p.184.
10.   Icher, *Building the Great Cathedrals*, p.184.
11.   Icher, *Building the Great Cathedrals*, pp.170, 174, 177.
12.   Windeatt, B. A. (trans.), *The Book of Margery Kempe* (Penguin Classics, 1994), p.96.
13.   Myers, A. R. (ed.), *English Historical Documents 1327–1485*, Vol. IV (Eyre & Spottiswood, 1969), p.1183.
14.   Harvey, J., 'The Education of the Medieval Architect', *Journal of the Royal Institute of British Architects*, June 1945, p.4.
15.   Icher, *Building the Great Cathedrals*, p.182.
16.   Knoop, D. & Jones, G. P., *The Mediaeval Mason, an Economic History of English Stone Building in the Later Middle Ages and Early Modern Times* (University of Manchester Press, 1933), p.115.
17.   Icher, *Building the Great Cathedrals*, p.165.
18.   Icher, *Building the Great Cathedrals*, p.166.
19.   Icher, *Building the Great Cathedrals*, p.185.

20. For example, Exodus 22:25, Leviticus 25:36, or Deuteronomy 23:19 (King James's Version).
21. Icher, *Building the Great Cathedrals*, p.166.

Chapter Seven: Ideas and people travelling

1. Frisch, T. G., *Gothic Art 1140–c.1450: Sources and Documents* (University of Toronto Press, 1987), p.76.
2. This whale-bone casket was named after a curator of the British Museum, Sir Augustus Woolaston Franks, not after the Frankish people.
3. Thurlby, M., *The Herefordshire School of Romanesque Architecture* (Logaston Press, 1999), p.6.
4. Sykes, P. M., *A History of Persia*, Vol 2 (Macmillan, 1915), p.191.
5. In fact, the war lasted more than a hundred years and the fighting was not continuous, so the title is misleading.
6. Lethaby, W. R., *Westminster Abbey: The King's Craftsmen* (Duckworth & Co., 1956) p.110.
7. Rowling, M., *Everyday Life in Medieval Times* (Batsford, 1968), p.157.
8. See Exodus 20.
9. Icher, F., *Building the Great Cathedrals* (Harry N. Abrams, 1998), p.64
10. Frisch, *Gothic Art 1140–c.1450*, p.6.
11. Tisdall, M. W., *God's Beasts* (Charlesfort Press, 1998), pp.73–4.
12. Acanthus symbolized eternal life because it was deemed to last forever. Indeed, as any gardener knows, once it is growing in a garden, it can be hard to get rid of it.

Chapter Eight: Patronage

1. *Southwell Minster Guide* (English Life Publications Ltd, 1999), p.25.
2. Field, J., *Kingdom, Power and Glory* (James & James, 1996), pp.17–18.
3. John, E., *Reassessing Anglo-Saxon England* (Manchester University Press, 1996), p.99.
4. *Christchurch Priory Guide*, p.18.
5. Hughes, J., 'Audley, Edmund (c.1439–1524)', *Oxford Dictionary of National Biography* Online at: www.oxford.dnb.com. Accessed 10 September 2017.
6. [AUTHOR TO SUPPLY (Audley cross-examining a Lollard child, folio 221)]
7. Froissart, J., *Chronicles*, G. Brereton (trans. & ed.), (Penguin, 1978), p.187.
8. *Cymbeline*, Act IV sc.ii, lines 333–4.
9. Jocelin of Brakelond, *Chronicle of the Abbey of Bury St Edmunds*, Greenaway, D., & Sayers, J., (trans. & ed.) (Oxford University Press, 1989), p.103.
10. *Sherborne Abbey Guide* (Friends of Sherborne Abbey, 2009), p.7.
11. Platt, C., *Medieval England: A Social History and Archaeology from the Conquest to 1600 AD* (Routledge, 1978), p.69.
12. Platt, *Medieval England*, p. 69.
13. Jocelin of Brakelond, *Chronicle of the Abbey of Bury St Edmunds*, p.103.

14. Mildenhall was a manor the abbey was trying to buy for its annual income, some of which went directly to the abbot. It was said to be worth £70 a year so the abbot offered five hundred marks to Richard I in 1190, but the king demanded a thousand. The money was found.

15. M.T. Clanchy, *England and its Rulers 1066–1272* (Blackwell, 1998), p.95.

16. Jocelin of Brakelond, *Chronicle of the Abbey of Bury St Edmunds*, p.103.

17. Green, L., *Building St Cuthbert's Shrine: Durham Cathedral and the Life of Prior Turgot* (Sacristy Press, 2013), p.51.

18. Green, *Building St Cuthbert's Shrine*, p. 51.

19. Green, *Building St Cuthbert's Shrine*, p.51.

20. Frisch, T. G., *Gothic Art 1140–c. 1450: Sources and Documents*, p.38.

Chapter Nine: How did they do it?

1. Shelby, L. R., *Gothic Design Techniques, the Fifteenth-Century Design Booklets of Mathes Roriczer and Hanns Schmuttermayer* (Southern Illinois University Press, 1997), pp.3–4.

2. Isidore of Seville, *The Etymologies*, Barney, S. A., Lewis, W. J., Beach, J. A. & Berghof, O. (trans.) (Cambridge University Press, 2010)

3. Salzman, L. F., *Building in England down to 1540*: *A Documentary History* (Clarendon Press, 1952), p.445.

4. Colvin, H. M (ed.), *The History of the King's Works*, Vol I (HMSO, 1963), pp.15–16.

5. Frisch, T. G., *Gothic Art 1140–c. 1450: Sources and Documents*, (University of Toronto Press, 1987) p.31

6. Frisch, *Gothic Art 1140–c. 1450*, p.31

7. Theophilus, *On Divers Arts*, Hawthorne, J.G. & Stanley Smith, C. (trans.), (Dover Publications Inc., 1979), Book II, Chapter 17.

8. Frisch, *Gothic Art 1140–c.1450*, pp.14–23.

9. Frisch, *Gothic Art 1140–c.1450*, p.18.

10. Frisch, *Gothic Art 1140–c.1450*, pp.22–3.

11. Barnes, C. F., Jr., *The Portfolio of Villard de Honnecourt: a new critical edition and facsimile* (Ashgate, 2009), pp.198–9.

12. Durandus, Guglielmus. *The Rationale Divinorum Officiorum* (Fons Vitae, 2007), pp.8–13.

13. An alternative interpretation of the number eight is that it represents the days from Palm Sunday to Easter Sunday inclusive. When seeing something octagonal, a Christian should reflect on the events of that extraordinary week ending, of course, with Christ's resurrection into new life.

14. Shelby, *Gothic Design Techniques, the Fifteenth-Century Design Booklets of Mathes Roriczer and Hanns Schmuttermayer*, which contains transcripts of the works of Roriczer and Schmuttermayer.

15. Shelby, *Gothic Design Techniques, the Fifteenth-Century Design Booklets of Mathes Roriczer and Hanns Schmuttermayer*, p.127.

16. Shelby, *Gothic Design Techniques, the Fifteenth-Century Design Booklets of Mathes Roriczer and Hanns Schmuttermayer*, p.28.

17. Shelby, *Gothic Design Techniques, the Fifteenth-Century Design Booklets of Mathes Roriczer and Hanns Schmuttermayer*, p.128.

18. Jestaz, B., *Architecture of the Renaissance from Brunelleschi to Palladio* (Thames & Hudson, 2010), p.143.

19. Jestaz, *Architecture of the Renaissance from Brunelleschi to Palladio*, pp.144–5.

20. Moorhouse, S., & Moorhouse, C., *Medieval Charcoal Burning* (Yorkshire Archaeological Society, undated), p.1.

21. Geddes, J., 'Iron', *English Medieval Industries*, in Blair, J. & Ramsay, N. (eds), *English Medieval Industries* (Hambledon, 2001), p.186.

22. Cherry, J., Medieval Goldsmiths (British Museum Press, 2011), p.38.

23. Salzman, *Building in England down to 1540*, p.175.

24. Erskine, A. M. (trans.), *The Accounts of the Fabric of Exeter Cathedral, 1279–1353* (Devon & Cornwall Record Society, 1983), Vol. 26, p.224.

BIBLIOGRAPHY

Ackerknecht, E.H., *A Short History of Medicine* (The Johns Hopkins University Press, 1955)

Adomnán of Iona, *Life of St Columba* (Penguin, 1995)

Anderson, M.D., *The Imagery of British Churches* (John Murray, 1955)

Anderson, M.D., *History and Imagery in British Churches* (John Murray, 1971)

Attwater, D., *A New Dictionary of Saints*, revised by Cumming, J. (Burns & Oates, 1993)

Augustine, St., *The Rule of Saint Augustine*, Van Bavel, T.J., (ed.) & Canning R. (trans.) (Image Books, 1986)

Barlow, F., *Thomas Becket* (Phoenix Giant, 1986)

Barnes, C.F., Jr, *The Portfolio of Villard de Honnecourt, a new critical edition and facsimile* (Ashgate, 2009)

Barnwell, P.S. & Pacey, A. (eds), *Who Built Beverley Minster?* (Spire Books, 2008)

Bartlett, R. (ed.), *Medieval Panorama* (Thames & Hudson, 2001)

Bartlett, R., *England under the Norman and Angevin Kings 1075–1225* (Clarendon Press, 2000)

Bede, *The Ecclesiastical History of the English People*, McClure, J. & Collins, R. (trans.) (Oxford University Press, 1969)

Bede, *The Lives of the Holy Abbots of Wearmouth and Jarrow*, Grocock, C. & Wood, I. N. (trans. & eds), (Oxford University Press, Oxford, 2013)

Bell, D. (ed.), *The Phaidon Encyclopaedia of Art and Artists* (Phaidon, 1978)

Benedict, St., *The Rule of St Benedict*, Abbot Parry OSB (trans.) (Gracewing, 1990)

Benedictow, O., *The Black Death 1346–1353: A Complete History* (BCA, 2004)

Bentley, J., *A Calendar of Saints: The Lives of the Principal Saints of the Christian Year* (Guild Publishing, 1986)

Biller, P. & Hudson, A. (eds), *Heresy and Literacy, 1000–1530* (Cambridge University Press, 1994)

Binding, G., *High Gothic: The Age of the Great Cathedrals* (Taschen, 1999)

Binding, G., *Medieval Building Techniques* (Tempus Publishing, 2003)

Binski, P., *Medieval Death: Ritual and Representation* (British Museum Press, 1996)

Blair, J. & Ramsey, N. (eds), *English Medieval Industries*, (Hambledon, 2001)

Blair, J., *The Church in Anglo-Saxon Society* (Oxford University Press, 2005)

Blatch, M., *In and Out of Churches, Part 1* (publisher unknown, 1990)

Blatch, M., *The Churches of Surrey* (Phillimore, 1997)

Boutflower, D.S. (trans. & ed.), *The Life of Ceolfrid, Abbot of the Monastery at Wearmouth and Jarrow* (Llanerch Press, 1912)

Bowyer-Smyth, V., *From Caen to Canterbury: The Stone in the Cathedral* (undated pamphlet)

Bradley, S.A.J. (trans. & ed.), *Anglo-Saxon Poetry* (Everyman,1981)

Bristow, J., *The Local Historian's Glossary of Words and Terms* (Countryside Books, 2001)

Brooke, C. N. L., *Churches and Churchmen in Medieval Europe* (Hambledon Press, 1999)

Brown, M.P., *The World of the Luttrell Psalter* (British Library, 2006)

Burckhardt, T., *Chartres and the Birth of the Cathedra*, (Golgonooza Press, 1995)

Butcher, A.F., 'The Origins of Romney Freemen, 1433–1512', *Economic History Review*, Vol. 27, 1974

Butler, A., *Lives of the Saints*, Walsh, M. (ed.) (Burns & Oates, 1985)

Campbell, J. (ed.). *The Anglo-Saxons* (Penguin, 1982)

Cannon, J., *Cathedral: The Great English Cathedrals and the World that made them* (Constable, 2007)

Carter, R.O.M. & Carter, H.M., 'The Foliate Head in England', *Folklore*, Vol. 77 (1966)

Cennini, Cennino d'Andrea, *The Craftman's Handbook*, Thompson, Jr., D.V. (trans.) (Dover Publications, 1960)

Cherry, J., *Medieval Goldsmiths* (British Museum Press, 2011)

Clanchy, M.T., *England and its Rulers 1066–1272* (Blackwell, 1998)

Clifton, C.S., *Encyclopedia of Heresies and Heretics* (ABC-Clio, 1992)

Coales, J., *The Earliest English Brasses: Patronage, Style and Workshops 1270–1350* (Monumental Brass Society, 1987)

Cobban, A., *English University Life in the Middle Ages* (UCL Press, 1999)

Coldstream, N., *Masons and Sculptors* (British Museum Press, 1991)

Coldstream, N., *Builders and Decorators: Medieval Craftsmen in Wales* (Cadw, 2008)

Cole, E. (ed.), *A Concise History of Architectural Styles* (A&C Black, 2003)

Colvin, H.M. (ed.), *The History of the King's Works*, Vol. I The Middle Ages (HMSO, 1963)

Colvin, H.M. (ed.), *Building Accounts of King Henry III* (Clarendon Press, 1971)

Cook, W.R. & Herzman, R.B., *The Medieval World View: An Introduction* (Oxford University Press, 2004)

Coulton, G.C., (trans. & ed.), *Life in the Middle Ages, Volume 1* (Cambridge University Press, 1930)

Cross, F.L. (ed.), *The Oxford Dictionary of the Christian Church* (Oxford University Press, 1958)

Crossley, D.W. (ed.), *Medieval Industry*, Council for British Archaeology Research Report No. 40 (Council for British Archaeology, 1981)

Crossley, F.W., *English Church Craftsmanship* (Batsford, 1941)

Davey, F., *William Wey: an English Pilgrim to Compostella in 1456* (Confraternity of St James, 2000)

Davis. R.H.C., *A History of Medieval Europe* (Pearson, 2006)

Douglas, D.C. (ed.), *English Historical Documents*, Vol. II (Eyre & Spottiswoode, 1969)

Duby, G., *The Age of the Cathedrals: Art & Society, 980–1420* (University of Chicago Press, 1981)

Duffy, E., 'The Parish, Piety, and Patronage in Late Medieval East Anglia: The Evidence of

Rood Screens', in French, K. L., Gibbs, G. G. & Kümin, B. A. (eds), *The Parish in English Life 1400–1600* (Manchester University Press, 1997)

Durandus, Guglielmus. *The Rationale Divinorum Officiorum: the Foundational Symbolism of the Early Church, its Structure, Decoration, Sacraments and Vestments, Books I, II, III and IV* (Fons Vitae, 2007)

Dyer, C., *Making a Living in the Middle Ages: The People of Britain 850–1520* (Yale University Press, 2002)

Dyer, C., *Standards of Living in the Later Middle Ages: Social Change in England c.1200–1520* (Cambridge University Press, 1989)

Eames, E., *English Tilers* (British Museum Press, 1992)

Einhard, *The Life of Charlemagne*, Thorpe, L. (trans.) (Penguin Books, 1969)

Eluère, C., *The Celts: First Masters of Europe* (Thames & Hudson, 1993)

Emery, G., *The Chester Guide* (Masons Design, 2003)

Erlande-Brandenburg, A., *Cathedrals and Castles: Building in the Middle Ages*, (Harry N. Abrams, 1995)

Erskine, A.M. (trans.), *The Accounts of the Fabric of Exeter Cathedral, 1279–1353* (Devon & Cornwall Record Society, 1983), Vol. 26

Evans, J. (ed.), *The Flowering of the Middle Ages* (Thames & Hudson, 1998)

Eveleigh Woodruff, C., 'The Financial Aspect of the Cult of St Thomas of Canterbury', *Archaeologia Cantiana*, Vol. 44, 1932

Farmer, D.H. (ed.), *The Age of Bede* (Penguin Books, 1965)

Farmer, D.H., *The Oxford Dictionary of Saints* (Oxford University Press, 1997)

Field, J., *Kingdom, Power and Glory* (James & James, 1996)

Finucane, R.C., *Miracles and Pilgrims: Popular Beliefs in Medieval England* (Book Club Associates, 1977)

Fox, P.A., 'Striving to Succeed in Late Medieval Canterbury: The Life of Thomas Folys, Publican, Mayor and Alderman *c.*1460–1535', *Archæologia Cantiana*, Vol. 129 (2009)

Frank, I.S., *A Concise History of the Medieval Church* (Continuum, 1996)

Friar, S., *A Companion to the English Parish Church* (Chancellor Press, 1996)

Frisch, T.G., *Gothic Art 1140–c.1450: Sources and Documents* (University of Toronto Press, 1987)

Froissart, J., *Chronicles*, Brereton, G. (trans. & ed.) (Penguin, 1978)

Gameson, R.G.G., 'thelwold, Benedictional of', in Lapidge, M., Blair, J., Keynes, S. & Scragg, D. (eds), *The Blackwell Encyclopaedia of Anglo-Saxon England* (Blackwell, 2001)

Geary, P.J., *Furta Sacra: Thefts of Relics in the Central Middle Ages* (Princeton University Press, 1978)

Gerson, P. (ed.), *The Pilgrim's Guide to Santiago de Compostela* (Harvey Miller, 1998)

Gimpel, J., *The Cathedral Builders* (Grove Press, 1961)

Goldberg, P.J.P., *Women in England c.1275–1525* (Manchester University Press, 1995)

Goldberg, P.J.P., *Women, Work, and Life Cycle in a Medieval Economy* (Clarendon Press, 1992)

Gombrich, E.H., *The Story of Art* (Phaidon, 1972)

Graham-Campbell, J., *Viking Art* (Thames & Hudson, 2013)

Grant, L., *Abbot Suger of St-Denis* (Longman, 1998)

Green, L., *Building St Cuthbert's Shrine: Durham Cathedral and the Life of Prior Turgot* (Sacristy Press, 2013)

Gregory of Tours, *The History of the Franks*, Thorpe, L. (trans.) (Penguin Books, 1974)

Grössinger, C., *The World Upside-Down: English Misericords* (Harvey Miller, 1997)

Hale, J., *The Civilization of Europe in the Renaissance* (HarperCollins, 1993)

Halsall, P. (ed.), *Abbot Suger: Life of King Louis the Fat* (1999). Online at: http://www.fordham.edu/halsall/basis/suger-louisthefat.html [accessed 10 May 2018]

Harvey, J., 'The Education of the Medieval Architect', *Journal of the Royal Institute of British Architects*, June 1945

Harvey, J., *English Mediaeval Architects: A Biographical Dictionary down to 1550* (Batsford, 1954)

Harvey, J., *The Gothic World 1100–1600: A Survey of Architecture and Art* (Batsford, 1950)

Henry of Huntingdon, *The History of the English People 1000–1154*, Greenaway, D. (trans.) (Oxford University Press, 1996)

Henson, D., *A Guide to Late Anglo-Saxon England from Ælfred to Eadgar: 871 to 1074 AD* (Anglo-Saxon Books, 1998)

Herlihy, D., *The Black Death and the Transformation of the West* (Harvard University Press, 1997)

Hindle, P., *Medieval Roads and Tracks* (Shire Archaeology, 1982)

Holt, R. & Rosser, G. (eds), *The Medieval Town: A Reader in English Urban History 1200–1540* (Longman, 1990)

Honour, H. & Fleming, J., *A World History of Art* (Laurence King, 1995)

Hopper, S., *To Be a Pilgrim: The Medieval Pilgrimage Experience* (Sutton Publishing, 2002)

Horrox, R. (trans. & ed.), *The Black Death* (Manchester University Press, 1994)

Howard, F.E., *The Mediæval Styles of the English Parish Church: A Survey of their Development, Design and Features* (Batsford, 1936)

Hugh of Poitiers, *The Vézelay Chronicle*, Scott, J. & Ward, J. O. (trans.) (Pegasus Paperbacks, 1992)

Hughes, J., 'Audley, Edmund (c.1349–1524)', *Oxford Dictionary of National Biography* (Oxford University Press, 2004)

Hunt, A., *Governance of the Consuming Passions: A History of Sumptuary Law* (MacMillan Press, 1996)

Hunt, T., *The Medieval Surgery* (Boydell, 1992)

Hutton, G. & Cook, O., *English Parish Churches* (Thames & Hudson, 1976)

Icher, F., *Building the Great Cathedrals* (Harry N. Abrams, 1998)

Isidore of Seville, *The Etymologies*, Barney, S. A., Lewis, W. J., Beach, J. A. & Berghof, O. (trans.) (Cambridge University Press, 2010)

Jacobus de Voragine, *The Golden Legend*, Stace, S.(trans.) (Penguin Books, 1998)

Jestaz, B., *Architecture of the Renaissance from Brunelleschi to Palladio*, Beamish, C. (trans.) (Thames & Hudson, 2010)

Jocelin of Brakelond, *Chronicle of the Abbey of Bury St Edmunds*, Greenway, D. &. Sayers, J. (trans.) (Oxford University Press, 1989)

John, E., *Reassessing Anglo-Saxon England* (Manchester University Press, 1996)

Jotischky, A. & Hull, C., *The Penguin Historical Atlas of the Medieval World* (Penguin Books, 2005)

Jusserand, J.J., *English Wayfaring Life in the XIVth Century*, Lucy-Smith, E. (trans.) (Putnam's Sons, 1931)

Keen, M. H., *England in the Later Middle Ages* (Routledge, 1973)

Keen, M. H., *English Society in the Later Middle Ages 1348–1500* (Penguin, 1990)

Kelly, J., *The Great Mortality: An Intimate History of the Black Death* (Fourth Estate, 2005)

Kelly. J.N.D., *Oxford Dictionary of Popes* (Oxford University Press, 1986)

Kendall, A., *Medieval Pilgrims* (Wayland Publishers, 1970)

King, E., *Medieval England from Hastings to Bosworth* (Tempus, 1988)

Knoop, D. & Jones, G.P., *The Mediaeval Mason: An Economic History of English Stone Building in the Later Middle Ages and Early Modern Times* (University of Manchester Press, 1933)

Kramer, S., *The English Craft Guilds and the Government* (AMS Press, 1905)

Kraus, H., *Gold was the Mortar: The Economics of Cathedral Building* (Routledge & Kegan Paul, 1979)

Jusserand, J.J., *English Wayfaring Life in the XIVth Century* (T Fisher Unwin, 1889; reprinted by Kessinger Publishing, 2003)

Lambert, M., *Medieval Heresy* (Blackwell, 1992)

Lapidge, M., Blair, J., Keynes, S. & Scragg, D. (eds), *The Blackwell Encyclopaedia of Anglo-Saxon England* (Blackwell, 2001)

Lethaby, W.R., *Westminster Abbey and the King's Craftsmen: A Study of Mediæval Building* (Duckworth & Co., 1956)

Lineham, P. & Nelson, J. L. (eds), *The Medieval World* (Routledge, 2001)

Loraine, H. R., *Knapton: Some Notes on the Church and the Manor* (Express Printing, 1986)

Loyn, H.R. & Percival, J., *The Reign of Charlemagne* (Edward Arnold, 1975)

Lucie-Smith, E., *The Thames and Hudson Dictionary of Art Terms* (Thames & Hudson, 1996)

Lysons, D. & Lysons, S., 'Antiquities: Ancient church architecture', *Magna Britannia: volume 4: Cumberland* (Cadell & Davies, 1816), pp. CLXXXIX–CCII. Online at: http://www. british-history.ac.uk/report.aspx?compid=50672 [accessed 10 May 2018]

MacCulloch, D., *A History of Christianity* (Allen Lane, 2009)

Manser, M. (ed.), *Dictionary of Saints* (Collins, 2004)

Mason, E., 'The Role of the English Parishioner, 1100–1500', *Journal of Ecclesiastical History*, Vol. 27 (1976)

McKisack, M., *The Fourteenth Century 1307–1399* (Oxford University Press, 1959)

McKitterick, R., *Medieval World* (Times Books, 2003)

Michael, M. A., *Stained Glass of Canterbury Cathedral* (Scala, 2004)

Moorhouse, S. & Moorhouse, C., *Medieval Charcoal Burning* (Yorkshire Archaeological Society, undated)

Mortlock, D. P. & Roberts, C. V., *The Guide to Norfolk Churches* (The Lutterworth Press, 2007)

Mortlock, D. P., *The Guide to Suffolk Churches* (The Lutterworth Press, 2009)

Myers, A. R. (ed.), *English Historical Documents*, Vol. IV (Eyre & Spottiswood, 1969)

Myers, A. R., *London in the Age of Chaucer* (University of Oklahoma Press, 1972)

Nilson, B., *Cathedral Shrines of Medieval England* (Boydell Press, 1998)

Norman, E., *The House of God: Church Architecture, Style and History* (Thames & Hudson, 2005)

Ohler, N., *The Medieval Traveller* (Boydell Press, 1989)

Orme, N., *Medieval Children* (Yale University Press, 2001)

Ormrod, M. & Lindley, P. (eds), *The Black Death in England* (Shaun Tyas, 1996)

Osborne, J., *Stained Glass in England* (Sutton Publishing, 1997)

Pacey, A., *Medieval Architectural Drawing* (Tempus Publishing, 2007)

Plagnieux, P., *Amiens: The Cathedral of Notre-Dame* (Monum Editions, 2005)

Platt, C., *King Death: The Black Death and its Aftermath in Late-Medieval England* (Routledge, 1996)

Platt, C., *Medieval England: A Social History and Archaeology from the Conquest to 1600 AD* (Routledge, 1978)

Platt, C., *The Atlas of Medieval Man* (St Martin's Press, 1979)

Postan, M.M. & Mathias, P. (eds), *The Cambridge Economic History, Vol. II: Trade and Industry in the Middle Ages* (Cambridge University Press, 1987)

Reed, P., *Church Architecture in Early Medieval Spain c.700–c.1100* (Shaun Tyas, 2016)

Remnant, G. L., *A Catalogue of Misericords in Great Britain* (Clarendon Press, 1969)

Riall, N., *Henry of Blois, Bishop of Winchester: A Patron of the Twelfth-Century Renaissance*, Hampshire Papers 5 (Hampshire Papers Committee, 1994)

Rigby, S.H., *English Society in the Later Middle Ages* (Macmillan, 1995)

Roberts, E., *The Wall Paintings of Saint Albans Abbey* (Fraternity of Friends of St Albans Abbey, 1993)

Rowling, M., *Everyday Life in Medieval Times* (Batsford, 1968)

Salzman, L.F., *Building in England down to 1540: A Documentary History* (Clarendon Press, 1952)

Santolaria Tura, A., *Glazing on White-Washed Tables* (Documenta Universitaria, 2014)

Saul, N. (ed.), *Age of Chivalry: Art & Society in Late Medieval England* (Brockhampton Press, 1992)

Schofield, J. & Vince, A., *Medieval Towns* (Leicester University Press, 1994)

Scott, J. & Gray, G., *Out of the Darkness* (Axminster Printing Company, undated)

Senderowitz Loengard, J. (ed.), 'Introduction', *London viewers and their certificates, 1508–1558: Certificates of the sworn viewers of the City of London* (1989), pp.11–65. Online at: http://www.british-history.ac.uk/report.aspx?compid=36052 [accessed 10 May 2018]

Shadwell, W. J., *A Handbook of Medieval Misericords*, http://content.yellowgrey.com/ms/a_handbook_of_medieval_misericords.php [accessed 10 May 2018]

Sharpe, R. (ed.), 'Folios ccxxii–ccxxx: Items relating to 1331–1334', *Calendar of Letter-books of the City of London*: E: 1314–1337 (His Majesty's Stationery Office, 1903), pp.262–71 http://www.british-history.ac.uk/report.aspx?compid=33117&strquery=ramsey [accessed 10 May 2018]

Shelby, L. R., 'Medieval Masons' Templates', *The Journal of Architectural Historians*, Vol. 2 (May 1971)

Shelby, L. R. (ed. & trans.), *Gothic Design Techniques: The Fifteenth-Century Design Booklets of Mathes Roriczer and Hanns Schuttermayer* (Southern Illinois University Press, 1997)

Sneesby, N., *Etheldreda, Princess, Queen, Abbess and Saint* (Fern House, 1999)

Southern, R.W., *The Making of the Middle Ages* (Hutchinson, 1967)

Sumption, J., *Pilgrimage: An Image of Mediaeval Religion* (Rowman & Littlefield, 1975)

Swanson, H., *Medieval British Towns* (Macmillan Press, 1999)

Swanson, H., *Building Craftsmen in Late Medieval York* (University of York, 1983)

Swanton, M. (trans. & ed.), *The Anglo-Saxon Chronicles* (Phoenix Press, 2000)

Sweetinburgh, S.M., *The Role of the Hospital in Medieval England* (Four Courts Press, 2004)

Sykes, P.M., *A History of Persia*, Vol. 2 (Macmillan, 1915)

Theophilus, *On Divers Arts*, Hawthorne, J. G. & Stanley Smith, C. (trans.) (Dover Publications, 1979)

Thurlby, M., *The Herefordshire School of Romanesque Sculpture* (Logaston Press, 1999)

Tisdall, M.W., *God's Beasts* (Charlesfort Press, 1998)

Toman, R. (ed.), *Gothic: Architecture, Sculpture, Painting* (Ullmann & Konemann, 2007)

Unwin, G., *The Guilds and Companies of London* (Frank Cass & Co., 1963)

Urry, W., *Thomas Becket: His Last Days* (Sutton Publishing, 1999)

Vasari, Giorgio, *Lives of the Artists*, Vol. I, Bull, G. (trans.) (Penguin, 1965)

Watkins, Dom Basil OSB (ed.), *The Book of Saints* (A&C Black, 2002)

Watson, P., *Building the Medieval Cathedrals* (Cambridge University Press, 1976)

Webb, D., *Pilgrimage in Medieval England* (Hambledon, 2000)

Weir, A., *Eleanor of Aquitaine, by the Wrath of God, Queen of England* (Jonathan Cape, 1999)

William of Malmesbury, *The Kings Before the Norman Conquest*, Stephenson, J. (trans.) (Seeleys, 1989)

Windeatt, B. A. (trans.), *The Book of Margery Kempe* (Penguin Classics, 1994)

Ziegler, P., *The Black Death* (Sutton Publishing, 1997)

INDEX